PIMLICO

22

THE DONKEYS

Alan Clark was educated at Eton and Christ Church, Oxford. He served in the Household Cavalry before qualifying for the Bar in 1955. In 1974 he became Conservative MP for Plymouth Sutton and went on to hold a number of ministerial posts. He wrote several works of military history: *The Fall of Crete, Barbarossa: The Russo-German Conflict 1941-45* and *Aces High: The War in the Air over the Western Front.* He also published his *Diaries.* Alan Clark died in 1999.

THE DONKEYS

ALAN CLARK

PIMLICO

PIMLICO

20 Vauxhall Bridge Road, London SW1V 2SA

London Melbourne Sydney Auckland Johannesburg
and agencies throughout the world

First published by Hutchinson 1961
Pimlico edition 1991
Reprinted 1993 (twice), 1994 (twice), 1995, 1996, 1997,
1998, 1999 (twice)

Printed and bound in Great Britain by
Mackays of Chatham plc, Chatham, Kent

ISBN 0-7126-5035-0

For

JANE

Contents

The photographs are reproduced by permission of the Imperial War Museum. The map on page 45 is adapted from the Ordnance Survey map in *The Official History of the War* by permission of the Controller of H.M. Stationery Office.

Ludendorff: 'The English soldiers fight like lions.'

Hoffman: 'True. But don't we know that they are lions led by donkeys.'

FALKENHAYN: *Memoirs*

Introductory Note

This is the story of the destruction of an army—the old professional army of the United Kingdom that always won the last battle, whose regiments had fought at Quebec, Corunna, in the Indies, were trained in musketry at Hythe, drilled on the parched earth of Chuddapore, and were machine-gunned, gassed and finally buried in 1915.

I was drawn to this subject almost by chance. While working in another field I came across the diary of an officer in the Leinsters and was overcome by the horror of the contents and the sense of resignation and duty that characterized the writing. I began serious research, back through the orders of battle and the unit records, in an effort to find out what happened to these men who endured for so long such incredible privations, such extremes of misery and squalor. Their casualties were frightful. In the first two hours of the Battle of Loos more British soldiers died than the total number of casualties in all three services on both sides on D-Day 1944. And slowly, as the field of operations widened, their fate became apparent. Again and again they were called upon to attempt the impossible, and in the end they were all killed. It was as simple as that.

My generation did not fight in the Second World War. To many of us the First is as remote as the Crimean, its causes and its personnel obscure and disreputable. I have tried to put down simply, factually, tediously even, what happened to these men in one year, 1915. Because in print they have no memorial. The huge cemeteries of regimented headstones that stand on 'ceded ground'—these are for the 'New Armies', the volunteers who died on the Somme the following year and for the conscripts slaughtered at Passchendaele. The graves of the soldiers killed in 1915 are harder to find: clusters of white crosses that stand where the men actually fell on the sites of the German redoubts or of the advanced dressing stations, often away from the roads, hidden in folds of the ground, signposted only by the fading green notices of the War Graves Commission. And in the same way the evidence of their fate is scattered among unit records, official histories, regimental magazines published years afterwards. Today there are very few visitors to the graveyards. The visitors' book at the Bois Carré cemetery at Loos contains only three English names for the whole of 1959. And so it is with the sources which, undisturbed for decades, gather dust in museum libraries.

I am anxious that this work should not be thought an 'indictment'. It is quite outside my intention to take part in arguments which relate, in any case, chiefly to the years of 1916 and 1917. This study is concerned simply with what the Army was ordered to do, and what happened when it attempted to carry out those orders; the results being important from a military-historical standpoint in that this year, 1915, saw the core of professional quality dissipated before it had been either properly equipped or substantially reinforced.

In compiling the material I owe an immense debt to that acknowledged master of military history, Captain B. H. Liddell Hart, who has allowed me access to his private files on the period and has been of the greatest help at every stage in the development of the book. I have also been greatly assisted by Miss Coombs of the Imperial War Museum Library, who has helped me in tracking down obscure items—often on the slenderest of leads. My thanks are also due to Captain Burgon Bickersteth, the historian of the Cavalry Brigade, for his help with documents and in conversation; to Colonel L. B. Beuttler for his assistance in extracting material from the War Office library; and to Captain G. C. Wynne from whose research on the period over the last thirty years, and from whose translations of German documents, I have drawn at length.

ALAN CLARK

Prelude: On the Aisne

Sir John French: The British Army will give battle on the line of the Condé Canal.
Sir Horace Smith-Dorrien: Do you mean take the offensive, or stand on the defensive?
Sir John French: Don't ask questions, do as you're told.[1]

In the first week of September 1914 the German armies to the east and north of Paris were in full retreat. The 'Battle' of the Marne—an engagement of manœuvre which, by the standards of later years, can be accounted almost bloodless—had broken the nerve of the German General Staff and they had authorized a general withdrawal to the line of the Aisne and the Chemin des Dames, intending, with their front thus shortened, to extend their right flank to the sea and make junction with the armies that had lately been investing Antwerp.

The war at this stage had a picturesque, traditional quality. The French infantry marched into action in their red-and-blue uniforms and felt caps, men and officers dressed like the pieces in 'l'Attaque', a parlour game lately introduced into Edwardian drawing-rooms,[2] whose very title suggested the one-sided trend of fashionable military thought. Photographs of those days show whole platoons of men forming up, under fire, in close order. Sometimes graced by a low farm wall, at others kneeling among the corn-stooks in a manner little different from that of the Old Guard at Waterloo, awaiting the signal to fire which their officer gave by dropping his sword, as at an execution. The

1. Brigadier-General Sir J. E. Edmonds, the official historian (Liddell Hart files).
2. 'L'Attaque' was just put on the market in 1890.

artillery clattered into action in the style of the military displays, or 'tattoos', that had followed an unaltered pattern for the last half-century: six-horse teams pulled the eighteen-pounders, or seventy-fives, and their limbers on spindly, iron-shod wheels, arraying them in exposed positions with a precision and geometric neatness that made counter-battery work by the enemy a simple problem of mathematics. The gunners had no protection other than a quarter-inch steel plate above the axle and the polished brasswork at breech and hub glinted in the September sunlight, often betraying their position before they had fired a round. The French Cuirassiers rode into action wearing full peacetime uniform with polished shakos and breastplates—'Dam'd fellows with their hair down their backs,' was the comment of General John Gough, Haig's Chief of Staff—and were mown down. Behind the lines was the excitable atmosphere of 'a prolonged field-day with a bit of circus thrown in for good measure'. Wealth and social distinction still counted for much, particularly when, as in all previous wars, it was both necessary and desirable to supplement the issued equipment by personal expenditure. 'You may hear at any time the sound of shot-guns and come across a party of officers shooting pheasants. There is a pack of beagles run by most cavalry units. . . .' Charteris, Haig's Chief of Intelligence, tells how he had '. . . a young Prince of Orleans attached to us as a sort of unofficial interpreter, also a French banker with a magnificent car. Both are very anxious to do anything for anybody.' (They seem to have spent most of their time driving English officers backwards and forwards from Paris.) His own staff consisted of '. . . a diamond merchant, an engineer from Vickers, and a brewer. The diamond merchant is appropriately rich; anyhow, he has placed at my disposal a very fine Rolls-Royce in which I can do my trips behind the lines.'

Many others found all this fuss objectionable and tried to ignore the war. The proprietor of the *Trois Tours* at Poperinghe was constantly complaining to all and sundry of the way in which the horses of the 9th Lancers were eating the bark off his trees—two years later the inn had been literally erased from the face of the earth and his orchard was a flat pool of mud. At least one of the grander restaurants in Paris excluded officers in uniform, and the sign '*Pas de chevaux*' was hung outside many of the châteaux in Picardy.

For at this early stage in the fighting the horses were everywhere. It was the cavalry, Queen of the battlefield since the Middle Ages, that caught the eye and the imagination: the Scots Greys, the 4th Hussars, the 5th Lancers, the 9th Lancers, the 12th Lancers, the 16th Lancers, the 18th Hussars, the 20th Hussars—in the Expeditionary Force it seemed that there were nearly as many regiments of horse as of foot.[1] In troop and squadron strength they trotted about the autumn countryside, pennants fluttering from the tips of their lances, young men and their grooms from fox-hunting families the length and breadth of Britain, eager for 'a go at the Boche'.

And this moment, of all, was their opportunity. Largely owing to the intervention of Galliéni and the famous 'Taxi-cab Army' the armies of Kluck and Bülow had become separated, one on each side of the Marne, connected by a front of some thirty miles that was almost without protection. Across this exposed flank streamed columns of transport and supply, and the whole confused mass of support echelons that were crowding up on each other as the fighting armies executed their turnabout. To protect this vulnerable and congested region the Germans were forced to rely on some scattered troops of Uhlans, two battalions of Jäger (sharpshooters on bicycles) and a few dismounted oddments that had straggled in. This motley force, without adequate central direction, had to hold the crossings of the Grand and Petit Morin rivers for a period—it could not be less than five days—while von Kluck withdrew from his salient.

Opposite them was the B.E.F. with its mass of cavalry

1. Actually the proportion was 18 cavalry to 78 infantry, but 17 cavalry regiments were in the line compared with only 42 infantry battalions.

and, under the command of cavalry officers, now offered what was to be, in Europe at any rate, the last and greatest opportunity in the history of the arm. A resolute thrust, pressed with even a semblance of the disregard for casualties that characterized later operations under the same commanders, would have broken through this screen and ruptured the enemy's supply lines. Kluck's army—which had been virtually without rations since the 5th September—would have been cut off and wholesale surrenders would have resulted.

But what followed was disappointing. The forward movement of the British cavalry was timid and hesitant. Sometimes, very rarely, the young officers had their hearts' desire and there were encounters with Uhlans or *Garde Dragoner*.[1] More often a Maxim, chattering elusively from some distant barn or copse, would cause a whole squadron to dismount and delay—perhaps until dusk.

Largely responsible for this faltering approach was Sir John French, the Commander-in-Chief. He was slow even to realize that the Germans had altered direction and on the 6th September his forces lost all effective contact with the enemy. This seems to have perplexed him, as did a request from Joffre, that evening, to push northward (i.e. straight into the gap, instead of north-west to relieve the imaginary pressure on Franchet D'Esperey's 5th French Army). In his diary that evening he wrote: 'It now became necessary to study the situation with great care,' and the following day: 'My intention to close at all speed with the enemy had to be tempered by consideration for the French Armies on my flank, both of which were opposed by much larger forces.'

Thus he was flatly ignoring the precepts laid down in the Field Service Regulations[2] that 'A Commander who has gained a strategical advantage may have to act at once in

1. A description of the charge of the 9th Lancers at Frétoy on the 7th September—probably the last encounter in which a member of the British Army was wounded by a *lance*—is given in Appendix No. 1.

2. Part 1, 1909 (Sec. 99.3).

order to prevent the enemy bringing about conditions more favourable to himself'—and that 'All pursuing troops should act with great boldness and be prepared to accept risks that would not be justifiable at other times.'

Attitudes at corps and divisional level were equally cautious and leisurely. General Sir Hubert le Poer Gough, the Commander of the 2nd Cavalry Division, went, on the 8th September, to a funeral which occupied him for the whole day and, on the 9th, did not get his cavalry out of their billets until three o'clock in the afternoon. On the 10th the whole divisional front was thrown into confusion because its leading echelon—the 5th Cavalry Brigade under Sir Philip, later Lord, Chetwode—executed a complete circle and came up in the evening against the left flank of its own support line—a manœuvre ascribed in the *Official History* as due to 'unfamiliarity with the terrain and the fatigued condition of horses and men'.

It is not surprising that under such conditions of leadership the English advance averaged somewhat less than seven miles a day. Of the two corps, Haig's went the slower owing to its commander's insistence on keeping his infantry in front, with the cavalry waiting behind—an unexpected order of battle from one who had written before the war that 'the role of cavalry on the battlefield will always go on increasing' and that 'the organization and training of cavalry must have as its basis the necessity of mass tactics'.

For a week the British cavalry meandered over the unfamiliar terrain disturbing the harvest, while among them the columns of infantry stopped and started and stopped again. The enemy was seldom seen. Gunfire was heard only occasionally. Then, on the 12th, when it was already too late, Sir John French ordered Haig to let his cavalry loose and 'get over the Aisne as soon as you can'.

'By now everyone was regarding the advance as a "pursuit". The roads were strewn with abandoned German equipment. I myself saw hundreds of lances left behind and there were many stragglers. The orders that "the crossings over the Aisne will be seized" were understood by all to mean that the advanced guards should push over the river, the main bodies remaining on the south bank. But everyone was very tired and, to put the lid on it, the 12th September provided a real wet afternoon and evening. In the afternoon heavy firing was heard south of Soissons and, although aviators had reported only a rearguard at Bazochs, ahead Haig ordered his divisions *not* to proceed as far as the Aisne but to halt ahead of the 1st Division at Vaucère and the 2nd at Dhuizel! His later excuse was that the 3rd Division on his left was not keeping up and the French 18th Corps on his right was also behind and could not be relied on. His usual Scots caution. As it happened, even the reduced march ordered by Haig was not completed until after dark.'[1]

But already the lava of the first eruption was hardening. The beginnings of trenches were being dug; strands of barbed wire were being staked out on the open ground; the last hours of 'fluid' warfare were ticking past as the front took shape in the form that it was to retain with but small variation for the next four years.

It was twenty years before this double failure—at both tactical and strategic level—was explored by students at the War Office, and even then their findings were given only a restricted circulation.[2] Among others, these points were given prominence:

'(1) *The necessity for orders to make their intention unmistakeable to the recipients.* None of the G.H.Q. orders at this time disclose the intention of the C. in C.

'(2) *The importance of not losing touch with the enemy.* For example, there was no reconnaissance in front of 1st Corps (Haig) on the night of the 13–14th September. As a result of this the advanced guards of both divisions were forced to

1. Sir J. E. Edmonds (L.H. files).
2. *Tour of the Aisne*, H.M.S.O., 1934. The italics are mine.—A.C.

deploy on the morning of 14th in the confusion of surprise, from cramped valleys about Troyon and La Maison Brulée, and found themselves crowded on ground too restricted for their proper employment and in complete ignorance of their surroundings.

'(3) There was a *hopeless lack of concentration.* The B.E.F. advanced to the Aisne with its divisions spread over the front and all were committed to battle on the 13th, though none were heavily engaged. There was no reserve except the 19th Infantry Brigade on the extreme left. No effort was made by G.H.Q. to concentrate superior force against the gap in the German lines and, having no reserve, G.H.Q. were unable to reinforce 1st Corps front on the 13th, where alone there was room for manœuvre.'

It is of interest to examine the conduct of the 'Battle' of the Aisne because of its significance as a background to the offensives of later years. In the following months, and continuing right up until the winter of 1917, the British commanders were to make every effort, spending the lives of their men with profligacy, to reproduce the sort of conditions of open warfare and 'cavalry country' that had confronted them on the Aisne in the autumn of 1914. But their handling of operations at this time gives no confidence that they would have been any more efficient or imaginative, had their wish been granted, than they were in coping with the siege-like conditions that set in after the first great opportunity was lost.

It may be suggested that in the preceding half-century the British commanders had acquired reputations that were greatly out of proportion to their achievements. Zulus, Afghans, Dervishes, Chinese—all these and even, in the end, Boers—had been defeated. But distance had magnified the severity of those 'struggles' and they had been still further exaggerated by the newspapers—themselves responding to the public appetite for glory during these long, tranquil years. Nor had it been inconvenient for the politicians to allow these inflated reputations to flourish. For the generals were far away; they could make no trouble; and their prowess, as it seemed, reflected glory on the home Government. Thus a popular tradition of heroic infallibility had been established which was to mate disastrously with the amateurish good humour and ignorance of contemporary military theory that was reality. For the adulation that had been their lot from Press and public had deluded the commanders with notions of their own ability and made them at the same time secure against dismissal by the politicians.

So it was that, as the leaves fell and the ground turned to mud and the German howitzers with their twelve-horse teams plodded patiently up to the line, the British Army was poised over an abyss. It could be saved only by a reckless squandering of the virtues which, like its delusions, sprang from a background of peace and a stable, ordered society. Bravery, perfect discipline, absolute conviction of right and wrong and the existence of God; a whole code of behaviour that is now little more than an object of derision—these were to be pitted against the largest and the most highly trained army in the world.

It could only be hoped that the British officers would profit rapidly from experience.

I

A Band of Brothers

> He [Sir John French] surrounds himself with capable leaders and staff officers, and not only brings his troops to a high degree of efficiency, but also makes his officers a band of brothers, and establishes a good comradeship between all arms and all ranks.
>
> *The Times*, on his appointment, 3rd August 1914

IN COMMAND was Sir John French. 'There was not a moment's hesitation about the appointment of Sir John,' said *The Times*, 'there was no painful canvassing of candidates, no acrimonious discussion, no odious comparison of the merits of respective generals, no hint of favouritism, of Party intrigue.' But to some it may have seemed unusual that these concepts were mentioned at all.

He was a weak-willed man of medium height, 'amiable enough', though 'petulant when thwarted'. He had 'a liking for the ladies', and rumour has it that this taste was not unconnected with his urgent need for £2,000 when he was commander of the Cavalry Brigade at Aldershot. He had borrowed the money from Douglas Haig, at that time his Brigade Major, and now one of his corps commanders.

For Haig, himself, having a superior officer in his debt was but one of a variety of fortuitous happenings that had so far compensated for a military talent which, although systematic, was not outstanding or original. He was not really a Haig of Bemersyde,[1] although he took the title on his

1. I have been asked by the solicitors to the Haig Trustees to set out the following facts concerning the late Field Marshal's lineage: 'He was directly descended from the 17th Laird and as such was entitled to bear the Quartered Arms. When he was born no Haigs were living at Bemersyde and the direct line through a younger son of the 17th Laird had petered out about 1840 when the last male heir died. Later about 1870 the sisters of that heir left the property

ennoblement, but came from the whisky-making side of the family. Hence he had not entered the Scots Greys, which might have been considered a natural choice, but had joined the 7th Hussars. He failed the Staff College examination. However, the Duke of Cambridge, who at that time had the right of nominating candidates was an acquaintance of Haig's elder sister, Henrietta. Under these auspices Haig applied a second time and the formality of an entrance examination was waived. From there he took frequent leave to attend shooting parties organized by his sister for the Prince of Wales, and these entries, boldly inscribed in the leave book, made their impression on his instructors.

None the less, in the final outdoor examination Haig did not shine and attracted unfavourable comment from General Plumer, who was conducting it. At thirty-eight he was still only a captain. During the Boer War Haig was Chief of Staff to French, who had command of the Cavalry Division. (He thought that 'the Boers were treated too generously') and afterwards he was made an A.D.C. to the King. From that time forward his ascent was more rapid and he became respected for his conventional opinions; as that 'Cavalry will have a larger sphere of action in future wars', and 'Artillery only seems likely to be really effective against raw troops'.

Finally, in 1905, he married the Hon. Dorothy Vivian, one of Queen Alexandra's maids of honour, and from then on his position in Court circles was unchallengeable.

An unfortunate result of the fact that his progress owed more to influential connections than to natural ability was that the Army seemed to contain many people who had tried to thwart Haig or who had, on account of superior quality, excelled him.

to a colateral who belonged to the same line as the Field Marshal, who owned Bemersyde until it was sold to the group of subscribers who bought it for the Field Marshal in 1921. During all this time the senior branch of the family had been living in America. In 1948 the Lord Lyon declared that in his opinion it was important for the Chief of the Family to be living in the family home, and so the Field Marshal's son was declared to be Haig of Bemersyde.'

For example General Grierson, his fellow corps commander in the B.E.F., had completely outmanœuvred Haig at the autumn exercises of 1912, to the embarrassment of all concerned, and to such an extent that the manœuvres had to be closed a day early. On arrival in France Grierson had died of a heart attack and Sir John French's choice as his successor was General Plumer, the erstwhile Staff College examiner who had taken such a poor view of Haig's performance. At the last moment, however, this decision was altered and Sir Horace Smith-Dorrien was appointed in his place.

As subordinates, too, there could be counted several divisional commanders enjoying only cool relations with Haig. One of these was General Allenby, commanding the 1st Cavalry Division who while at the Staff College had been preferred, when in direct competition with Haig, for the mastership of the Drag.

Perhaps it is on account of a feeling of insecurity engendered by this background that Haig's diaries, which are filled with information about his colleagues, seem unusually critical. For it is almost impossible to find mention of one whose abilities do not fill him with misgiving—unless it be Major-General Lomax, 'an experienced and practical leader, *most loyal to me*'.[1]

Loyalty, however highly he may have regarded it in subordinates, was not a virtue that Haig himself exhibited in relation to those above him. On the 11th August he cornered the King, who had been conducting a farewell review of the Expeditionary Force at Aldershot, and told him:

'. . . as I felt it my duty to do, that from my experience with Sir John in the South African War he was certain to do his utmost loyally to carry out any orders which the Government might give him. I had grave doubts, however, whether either his temper was sufficiently even, or his

1. The italics are Haig's. The implication shall be left for the reader to estimate—A.C.

military knowledge sufficiently thorough, to enable him to discharge properly the very difficult duties which would devolve upon him during the coming operations.'

The King's response to this disclosure is not recorded. Perhaps it was a little disappointing for Haig wrote further on that 'the King did not give me the impression that he fully realized the grave issues both for our country as well as for his own House, which were about to be put to the test', although (poor man) he '. . . *seemed anxious*'.

Haig's doubts about French extended to his Chief of Staff, Major-General Sir Archibald Murray:

'I had a poor opinion of his qualifications as a General. In some respects he seemed to me to be "an old woman". For example, in his dealings with Sir John. When his own better judgement told him that something which the latter wished put in Orders was quite unsound, instead of frankly acknowledging his disagreement,[1] he would weakly acquiesce in order to avoid an outbreak of temper and a scene.'

Haig added:

'However, I am determined to be thoroughly loyal and do my duty as a subordinate should, trying all the time to see Sir John's good qualities and not his weak ones, though neither of them [French and Murray] is at all fitted for the appointment which he now holds at this moment of crisis.'

Of Monro, who took Murray's place as G.O.C. 2nd Division, Haig wrote:

'Monro proved himself to be a good regimental officer and an excellent commander of the Hythe school of musketry,

1. i.e. Showing 'disloyalty'? Further evidence that Haig's attitude in matters of this kind was strictly subjective is offered by his entry regarding a slight disagreement with his own Chief of Staff, Brigadier-General John Gough: 'It was during the retreat on the night after the action at Villiers-Cotterets. After dinner at Mareuil, he, in his impetuous way, grumbled at my going on "retreating and retreating". As a number of the Staff were present, I turned on him rather sharply, and said that retreat was the only thing to save the Army, and that it was *his duty to support me* instead of criticizing. He was very sorry, poor fellow.'—*The Private Papers of Douglas Haig*, p. 87.

but some years with the territorials has resulted in his becoming *rather fat*. He lacks practical experience in commanding a division.'

In considering the younger officers Haig could find little grounds for satisfaction. Of the staff of Sir Henry Rawlinson, Bt., commander of the 4th Infantry Division, he wrote:

'His general staff consists of two Regulars, R. A. K. Montgomery, R.A., and Dallas, who had a bad sunstroke in India, from the War Office. Toby Rawlinson (his brother) acts as Mess President. He is now graded as Colonel, though he left the 17th Lancers as a Subaltern. Joe Laycock and the Duke of Westminster were A.D.C.s. There were two or three other officers about, who in peacetime were connected with motors or polo ponies.'

Of the Indian Corps:

'I felt surprise at the air of dejection and despondency which met me all round their headquarters, both outside, where orderlies and others were hanging about numbed with cold, and inside, where all ranks, staff officers, British and Native clerks seemed to be working together in three or four rooms on the ground floor. All the windows were shut and the atmosphere was, of course, very close. I came away feeling that things were not altogether in an efficient state in the Indian Corps.'

On visits to other units he found further grounds for complaint.

'I motored over in the morning to see the 17th Lancers. They gave me a great lunch in the Marie. The regiment is messing by squadrons. This may do very well at first, but in my opinion the officers of a regiment should always mess together in a "Regimental Mess" whenever possible.'

Nor were good manners any surer a road to Haig's favour: he described D'Urbal, commander of the French Army on his left, as 'a tall suave, elderly gentleman—rather an actor, the type of man seen on the stage playing the part of "the respectable Uncle"—and unpleasantly *polite*'.

While Haig could find little consolation as he looked around and beneath him, his superior was haunted by fear of the Secretary of State for War, Lord Kitchener.

Relations between Kitchener and French had been poor since the time of the South African War. Now they were worsened by Kitchener's reluctance to accept the restraints of a political, civilian, appointment. As the premier soldier of the Empire it seemed to him intolerable that he should not be allowed to give Sir John direct orders, as to a subordinate. He discussed frequently with his friends the possibility of assuming the post of Captain-General, or Generalissimo, and of holding it in addition to his existing office of Secretary of State, in order to formalize the responsibility which he bore for the supreme direction of British military strategy. When he travelled to France he always wore his uniform, and tricky points of protocol—always a fertile source of dispute among soldiers—would crop up.

The position was aggravated by Henry Wilson, nominally sub-Chief of Staff, who held an ill-defined but highly influential position of liaison with French General Headquarters, or 'G.Q.G.'. Henry Wilson's importance was twofold. In the first place he was a convinced 'Westerner', that is to say that he was deeply committed to the doctrine that the whole war effort of the Empire should be applied exclusively in Flanders[1] to the exclusion of all other theatres such as the Balkans and the Middle East. Second, he enjoyed the best possible relations with the French commanders—unlike many of his compatriots whose careers were equally involved—and from Foch and Joffre he would pick up a

1. In his diary he tells how he 'Dined with the King. Also Prince of Wales and Stamfordham [Private secretary to the King]. Had little talk with the King, but much with S who said among other things that I was more responsible for England joining the war than any other man. I think this is true' (*Memoirs of Sir Henry Wilson*, ed. Sir E. Callwell, 189).

variety of confidential information and political tit-bits often in advance even of their arrival at the Cabinet Room in London.

Wilson was convinced of his own indispensability, but power—real power, that is, as compared with mere leverage for intrigue—seemed to be eluding him. This may explain his morally tortuous behaviour. As Director of Military Operations at the War Office he had played a prominent role at the time of the scandalous 'Mutiny at the Curragh'[1] when he communicated Cabinet secrets to the Opposition and seems, as his diary shows, to have seen nothing improper in giving advice and encouragement, based on confidential information, to the Ulster Volunteers. He was friendly with Sir John French, they had been on intimate terms since the Curragh incident (which had forced French's, though not Wilson's, resignation) and the two men were joined in a mutual dislike of Lord Kitchener.[2] The Commander-in-Chief was less articulate in his expressions of dislike, but he seems to have listened happily enough as Wilson told him that the Secretary of State was as great an enemy of the B.E.F. as Moltke or Falkenhayn. Wilson also complained

1. In essence the Curragh affair (March 1914) centred round the stated preference of Brigadier-General Hubert Gough and fifty-one out of seventy officers of the 3rd Cavalry Brigade for 'dismissal' rather than action should they be ordered to enforce the Liberal Government's home rule policy against the protestant north of Ireland. A muddled War Office directive and the incompetent handling of the incident by Sir Arthur Paget, G.O.C. Ireland, obscured the real issue. This lay in the unwelcome irruption of the Army into domestic politics, and the establishment of an *entente* between its higher echelons and the Conservative opposition—an association of which there were to be several reminders during 1915.

2. Kitchener's biographer, Sir Philip Magnus, explains the background to this hostility. In July 1909 Kitchener had visited the Staff College at Camberley at the end of Wilson's period of service as Commandant. Kitchener had then questioned some aspects of Wilson's teaching, and Wilson had not replied with becoming modesty. Relying upon the licence which his social popularity had earned in many quarters, he had displayed a casual breeziness which the Field Marshal had deemed unsuitable. Five years later, on the 7th August 1914, Kitchener had occasion to summon Wilson to his room in the War Office in order to rebuke him for indiscreet discussion in Mayfair drawing-rooms of the transport arrangements of the B.E.F. Wilson hotly resented that rebuke, and thereafter did his best to poison the receptive mind of Sir John French against Kitchener.

constantly about Kitchener's policy of keeping a number of trained officers and N.C.O.s in England to serve as instructors for the new Kitchener's Armies that were being formed. (Actually, not enough were kept back.) Wilson professed to believe that Kitchener was mad: and he joked about Kitchener's 'shadow armies for shadow campaigns at unknown and distant dates'.

This friction between French and Kitchener had unfortunate side effects—not least of these being that produced on the running of French's own headquarters. When after three months it became necessary to replace Sir Archibald Murray (French's Chief of Staff), there developed a positive turmoil of intrigue and mortification. When the matter was first raised it was taken for granted that Wilson would succeed. On the 19th December he wrote in his diary: 'Saw Sir John twice this morning and again this evening. He talked as though it were settled that I was to be C. of S.' But four days later there was disquieting news:

'Sir John began by saying that he would speak very openly. He said no man had ever given another more loyal and valuable help than I had given him. He said that so long as he was alive and had power my future and my promotion were assured. He went on in this way for some time and then came to the real point. He said the Government and Kitchener were very hostile to me. They said my appointment would be very repugnant to the Cabinet and would shake confidence in the Army!'

It is evident that Wilson was not content to leave the protection of his interests in Sir John's hands, in spite of the latter's protestations. On Boxing Day he drove over to G.Q.G. at Chantilly with Huguet[1] and they discussed the whole question in the car. On arrival they recounted the state of affairs to Joffre in the presence of Delcassé[2] who, by chance, was also there.

1. French liaison officer at G.H.Q. at St. Omer.
2. French Foreign Minister.

'On this Delcassé said he would see Bertie[1] at once, and that if this interview was not satisfactory he would go over and see Asquith, that it was intolerable that I should be ruled out for policital reasons. . . . Delcassé had sent for Bertie and was crossing to England tonight, so it looks as though he were moving.'

Fortified by this, Wilson had another meeting with Sir John at which he said that: '. . . in my judgement he *must* remove Murray. He *must* beat Asquith on the matter of principle, and he *must* offer me the appointment.' Wilson declared that when offered the post he would refuse it. '. . . I can do no more than refuse the appointment that I have worked for and dreamed of for years.' And he went on to assure the Commander-in-Chief (though on what evidence is not clear) that Murray himself was anxious to vacate the post: '. . . five minutes after he is told that he is going to be given a Corps, he will thank God that the strain is over'. French's reaction to this harangue is not specifically recorded. It is possible that from experience he felt that Wilson might after all be prevented by his conscience from refusing the post when it was thus formally offered to him. At the time he was no more than 'charming and grateful', and gave Wilson to understand that he 'I think, perhaps, will use this loophole'.[2]

For a few days it seemed as if a compromise, and a typically unsatisfactory one, was to be the result of all this agitation; namely the retention of Murray in his position and of Wilson, and the other candidates, in theirs. G.Q.G. were keeping a watchful eye on the situation, and on 5th January Foch was able to write to Joffre and tell him that:[3]

'My telegram in cipher despatched today gave you a brief account of my knowledge of Field Marshal French's

1. Viscount Bertie, British Ambassador in Paris.
2. All quotations are taken, unless otherwise stated, from Sir E. Callwell's edition of Henry Wilson's *Memoirs*.
3. Liddell Hart, *Foch*.

intentions with regard to a prospective change in his staff. Whether he has asked to keep him [Murray] I don't know; I do not think so. But he may have abstained from asking for his recall; for I know that when he learned of the steps we had taken he said that in those circumstances he could do nothing.

'English pride demands that Murray stays where he is. Anyhow, Lord Kitchener and Mr. Asquith would not hear of General Wilson as his successor. When Murray is recalled and Wilson has gained people's confidence, I believe that we shall be able to progress. . . .'

In the meantime Wilson's hopes had been raised again when French, who was preparing his headquarters for a visit from Asquith and Kitchener '. . . leant over to me and said to me in a whisper: "You are such a brute, you will never be nice to people you don't like. Now I am going to get Asquith out here, why don't you make love to him?" By which he means that he still wants to have me as Chief of Staff.'

However, Foch's hopes of his protégé 'gaining people's confidence' showed little understanding of Wilson's character. The following week he was back in London and saw, among others, Lord Lansdowne, with whom he was characteristically indiscreet:

'I spoke freely about our relations with the French, and my proposal that they should send some representative men to see what Kitchener was doing, also of the strained relations between Sir John and "K"—also of my suspicions of Winston's intrigues—and so forth. He was charming, as usual.'

When he got back to France Wilson made contact with Robertson, at that time Quartermaster General and his leading opponent as candidate for the post. They went for a drive in Wilson's Rolls and Robertson said that though '. . . the offer was a tempting one, as it meant an increase of pay as well as of position, I did not wish to accept it'.[1]

It did seem that for the time being the whole matter was dying down, and Wilson left for a tour of the French line at the invitation of Joffre: 'No other officer in any army, not even a Russian, has been allowed to go down the French line except me.' Certainly it started off pleasantly enough. He was 'everywhere met by the Generals, who took the greatest pleasure in showing me things and making me as comfortable as they could. I went to Amiens, Chantilly, Chagny, Villiers Cotterets, Reims, Epernay, Bar-le-Duc, Remiremont. . . .' But at Remiremont there was bad news. A message '*en clair*' from Robertson writing *as Chief of Staff*, announcing the relief of Murray on grounds of ill-health, and Wilson's formal appointment as liaison officer with the French Army.

Wilson hurried back to St. Omer where he saw Sir John who, in some embarrassment it may be thought, repeated that 'nothing he could ever do for me for the work I had done could be enough, and that, so long as he held power, etc., etc.'. French did, however, agree to Wilson's immediate promotion to lieutenant-general (although he had no power to do so without reference back to the War Office) and there was talk of a K.C.B. But this latter hope turned out to be ill-founded; in the days immediately preceding publication it got around that Wilson's name was not, after all, on the list. French approached Kitchener, who replied that nothing could now be done as the King had signed the list. When the *Honours Gazette* appeared it was found that Wilson had been granted his temporary lieutenant-generalship as an 'honour'. Wilson ('The fools have given me another opening') at once wrote to Robb, the Military Secretary, claiming that this gave him 'permanent date', i.e. not a temporary rank at all—'but happily,' as his biographer says, 'this question never had to be put to the test'.

Sir John French seems to have had the idea of compensating for the only mediocre support that he had lent to

1. Sir William Robertson, *From Private to Field Marshal*.

Henry Wilson's aspirations by the bad manners with which he treated Robertson, the Government nominee. He ignored Robertson socially, insulted him in public on a variety of occasions, refused to mess with him and continued to sit next to Henry Wilson at meals. Robertson got his own back when Wilson tried to dabble in Staff matters by stalling him on one pretext or another, refusing him access to documents or sending them up days late. Naturally the smooth running of the British G.H.Q. was affected. Haig describes a characteristic incident:

'I went to Hazebrouck at 11.30 a.m. to see Sir John French. When I was shown into his room, Sir William Robertson (C.G.S.) followed. Sir John said would he kindly wait as he had something to say to me alone. Then when Robertson had gone he said that he had "nothing private to say, only he wished to make it clear to R. that he (F.) meant to see his army commanders alone occasionally, because R. had tried to insist that F. should not see any of his subordinate commanders unless he (R.) was present as C.G.S."!'

Certainly from contemporary documents one does not draw a reassuring picture of happy personal relations in the higher echelons of the B.E.F. Haig describes the Mess at G.H.Q. on a typical evening.[1]

'I motored to St. Omer and dined with Sir John French. Lieutenant-General H. Wilson was also dining. Brinsley Fitzgerald told me that the C.-in-C. had asked Wilson to join his Mess—a very great mistake, we both agreed, because he is such a terrible intriguer and is sure to make mischief. Wilson's face now looks so deceitful. By having W. in his Mess, while Robertson (the Chief of Staff) is only able to see him at stated times, the Commander-in-Chief is courting trouble. Billy Lambton (the Mil. Sec.) is weak, and quite under the influence of Wilson, it seems. Luckily,

1. Haig's diary for 12th March 1915.

Lambton is stupid, and more than once has unconsciously given away what H. Wilson has been scheming for.'

Personal rivalries at G.H.Q. were complicated at every stage by the intervention of the French whenever it was felt by them that the influence of 'dooble-Vay', as Wilson was known, was threatened. The first instance of this, and one which, it may be throught, provides the key to the whole situation, had arisen at the Dunkirk conference of November 1914:

'In imagined privacy Kitchener mooted his intention of recalling Sir John French and replacing him by Sir Ian Hamilton. Joffre and Foch had thought of asking that French should be replaced by Henry Wilson but they were not favourable to a change to the unknown which might weaken their existing influence over the British command. The following day Foch told Wilson privately of Kitchener's proposal, and suggested that French himself ought to be told. Next day, according to Wilson's diary: "Sir John and I went to Cassel at 3 p.m. when Sir John thanked Foch personally and in the warmest terms for his comradeship and loyalty. They shook hands on it, and the two parted great friends." Through this breach of confidence French and his staff were able to take steps both at home and in France to nullify the proposal. Also there is little doubt that some members of his staff took the shrewd course of informing Joffre, quite untruthfully, that Sir Ian Hamilton spoke the French language even worse than Sir John French. It is needless to emphasize the effect of this hint at G.Q.G. where the inter-allied situation in Flanders was already compared, with caustic humour, to The Tower of Babel.'[1]

Foch's disclosure naturally strengthened his influence over French, if it did not increase his respect for him. It is clear that Foch gauged aptly the character of the British Commander-in-Chief, if also that admiration for it was not the

1. Liddell Hart, *Foch*, p. 149.

reason for combating his recall. For, earlier, when Huguet (the French liaison officer with the British) told Foch that Sir John was aggrieved with him, he jocularly replied: 'Bah! It is of no importance; you have only to tell him that he has just saved England; that will put him in good humour again!' On this message being conveyed to French he made the gratified, though not immodest, retort: 'But, my dear fellow, I know it only too well, I knew it from the beginning.'[1]

Now, after Robertson's appointment, Foch wrote to Joffre:

'General Murray is leaving the English Army, ostensibly because of his health. Murray is replaced by General Robertson, a good choice in default of Wilson. Wilson remains head of operations and relations with us. His status has not been raised, but his position is growing more important.'

Lest these overt intrusions into the domestic politics of the British Army should have left too many ruffled coats, Foch added: 'It might be advisable to make another distribution of medals among the English generals. The Field Marshal [i.e. Sir John French] wished to remind me of it. . . .'[2]

1. Liddell Hart, *Foch*, p. 149.
2. *Ibid*, p. 152.

2

Winter in the Trenches

The region of the Lys basin and the plain of Flanders consists entirely of low-lying meadow. Throughout the winter months the clayey subsoil holds the water approximately two feet below the surface and there is a tendency for any minor declivity, whether natural or artificial, to become water-logged.

Introductory note to *War Office Manual*, 1913

NO-MAN'S-LAND was a grassy tufted waste, pock-marked with brown craters, with here and there the the stumps of broken trees and little greyish mounds which, from their situation and contour, suggested human origin. At a distance—it varied from 80 to 200 yards—stood the German emplacements. Through half-closed eyes, or when veiled by the damp mist that rose from the ground at dawn, the irregular line of grey and fawn hummocks that was the enemy breastwork might have been a stretch of dunes on the seashore, with the dark bundles of wire straggling from their lower reaches like wild blackthorn.

Sometimes, at night, it was absolutely still for minutes at a time. The voices of the enemy could be heard and even the click of a sentry's heels at inspection. A subaltern in the Black Watch wrote in his diary:

'I could hear some Boche playing Schubert; it was "The Trout", that bit that goes up and down, on an old piano. They must have got it in a forward dug-out; even so it was incredible how clear the sound came across. But before he got to the end someone put a flare up over Auchy and the whole of No-Man's-Land went pale green. A nervous sentry fired a short M.G. burst and firing started up all along the line. It went on and off for about half an hour. I never heard the pianist again, although Corporal Duffy said he was performing on the following night. I often wondered whether he survived the War.'

As the November fighting died down the British troops had found themselves holding a 'line' of scattered trenches, the majority of them scratched hastily in the soil while the battles were at their height, unconnected with each other and without any proper system of communication and support to the rear. When these were linked up into one continuous strip by the engineers many weaknesses became apparent: the successive counter-attacks that had been made in the last days of November had recovered much of the ground lost but, owing to the exhaustion of the men and their depleted numbers, had been brought up short before any enemy positions of natural strength. The instructions from G.H.Q. that not an inch was to be yielded, and the terrible cost at which the ground had been re-won, alike made it difficult to alter the line where this might have meant giving up even a few hundred yards of territory. And so the British front, like the last few inches of a high tide, was everywhere indented by little areas of high ground, or groups of buildings at road junctions, or other sorts of positions that offered unusual advantages to the enemy.

In this way whole stretches were subject to crippling enfilade fire from the German positions, that gave rise to a constant drain of casualties in holding on to them; the digging of communication trenches was particularly dangerous in sectors such as these and in some cases had to be abandoned altogether, which in turn meant that long frontages were without proper connection to the support areas and were dependent for the supply of ammunition and other essentials, and for the evacuation of wounded, the provision of reliefs and so forth, on the hazardous and uncertain night traffic along the fire-trench itself.

The trenches themselves were pitiful affairs. The infantry 'showed considerable lethargy and a marked disinclination to dig',[1] largely on account of the unfamiliarity of the medium and the G.H.Q. policy of switching units from sector to sector. This meant that the troops were seldom in a position long enough to effect any marked improvement, and there was a feeling that they were simply doing the work for those that came after, with the certainty that in the stretch where they themselves were next posted they would have to start all over again. This attitude persisted for many weeks, until it gradually became obvious that the condition of trench warfare was a permanent one. Certain regiments also, notably the Royal Scots and the Somerset, began to make it a point of prestige that 'no unit should ever have cause to complain when it takes over a stretch of line from us',[2] and with the spreading of this practice the strength and habitability of the line began to increase. Even when the will was there, however, there was a painful shortage of means.

Picks and shovels were considered plentiful when there were as few as two or three per platoon and efforts to commandeer them from civilian sources met with little success, as the Flemish peasants used to bury them rather than part with the tools of their livelihood. There was also a serious shortage of actual construction material, and particularly of sandbags and wattling for 'reveting' the sides of the trenches. The scarcity of sandbags was particularly serious in low-lying areas such as the Ypres salient and opposite Festubert, where the trenches were almost permanently waterlogged throughout the winter. In places such as these it was necessary for protection to construct a raised breastwork which, if it was adequately to protect against machine-gun fire, particularly at the very close ranges that separated

1. O.H., 1915, I, 28.
2. Ewing, *The Royal Scots*.

the troops in many areas, had to be at least eight feet thick.

'Sergeant Doherty was killed by a sniper while supervising a building fatigue. This is the eighteenth casualty and the fourth N.C.O. we have lost in this way since we came into the line on Tuesday—it is a frustrating business. The Boche has got perfect observation of our lines from Frezenburg Ridge. The snipers pick the men off in the evening before they can get started. We slave away all night building a parapet of loose earth—I have hardly seen a sandbag since our arrival—then in the morning he calls over a few "crumps" and they blow the whole thing to blazes, usually burying some poor wretches alive at the same time as they unearth a lot of dead ones!'

Throughout these bleak months the German artillery dominated the situation, making life a misery for the British troops who were obliged to hold the line in greater strength either than the enemy or the French, owing to their own shortage of guns. For whereas their allies could afford to make their front positions little more than outposts that could call up an immense weight of artillery fire at the least sign of any suspicious activity on the part of the enemy, the British were dependent on rifle fire to cope with marauding patrols and local attacks. This was due to two things: in the first place the eighteen-pounders used for direct support were few in number,[1] and hesitated to expose themselves except in an emergency owing to the fact that the heavier guns needed to support them against German counter-battery fire were almost entirely absent. Secondly, they were so starved of ammunition as to make it futile to reveal their position for the sake of throwing the meagre daily 'ration' of shells at the enemy. In actual fact for the entire B.E.F there was in the field only about three-fifths of the regulation

1. The regulation number of batteries per division—not always achieved—was obtained by reducing the number of guns from six to four per battery (O.H., 1915, I, 9).

amount *calculated on the experience of the Boer War*, and really little more than a day's supply in modern battle.[1]

The fire-power of the men in the front line was also seriously diminished by the shortage—amounting in cases to non-existence—of trench-mortars and hand-grenades. Of the latter a number of extemporized missiles were tried out, the most notorious being the 'jam-pot', the 'Battye bomb' and the 'hairbrush'. These were dangerous and difficult to construct, their ignition was chancy and impossible in wet weather, and in general it is likely that they caused as many casualties among the British as among the enemy. No 'Mills' hand-grenades were produced until the spring of 1915—by March only forty-eight had been delivered. The trench-mortar, an ultra-short-range howitzer, more or less portable, with which the Germans were making great destruction, was even more rare in the Expeditionary Force. One officer, however, managed to do a private deal with the French, paying cash for a number of old Coehorn siege-mortars which were found to bear the cypher of Louis Philippe!

. . . . .

It was in these conditions, starved of the equipment necessary in trench warfare, with little pretence even of artillery support and seriously short of trained junior officers and N.C.O.s, that the British troops were crowded into the fire-trenches to suffer throughout the winter months extremes of physical privation.

It rained incessantly. From the 25th October until the 10th March there were only eighteen dry days, and on eleven of these the temperature was below freezing. The trenches themselves became little less than culverts, replacing in rudimentary fashion the drainage system of the

1. For example figures for the 17th November (2nd Corps) are 3rd Division, 363 rounds per field gun, 5th Division, 323. Reserve in park, 6, 28. 3rd Corps, 45, 551 for all divisions—i.e. rather less than 300 per gun. *War Establishments*, Part I, p. 5 lays down minima of 528 and 280 for field-guns and howitzers, with a further 472 and 520 on lines of communication, in addition to the general reserve.

countryside which had been dislocated by the digging and artillery fire. It was impossible to dig deeper than eighteen inches without finding water, and along whole stretches of the line garrisons had to do their stint with the water waist-high, for the fire-step had crumbled away and there were not the materials to construct an adequate breastwork after the German fashion. Duck-boards were unknown and the wounded who collapsed into the slime would often drown, unnoticed in the heat of some local engagement, and lie concealed for days until their bodies, porous from decomposition, would rise once again to the surface. When the German guns opened fire the troops could only cower in the water because the dug-outs, built for protection during a bombardment, were themselves awash to roof level and stank intolerably from the dead that floated there.

In an effort to alleviate these conditions, relieving the men every twelve hours was tried (the German rota was four days in the line, two in support and four at rest), but this led to great administrative confusion, particularly in the immediate rear and over the allotment of billets themselves 'filthy and inadequate',[1] and to heavy casualties from sniping and shrapnel over the continuous traffic along the communication trenches.

It is thus not surprising that the 'wastage' from illness was very high—the more so in view of the fact that there were no proper facilities for drying clothing and the men frequently had to return to the line in the same soaking garments in which they had quitted it. 'De-lousing' stations were established, but the process consisted of no more than running a hot flat-iron over the troops' undergarments about once every ten days. Although the strictest criteria were applied before men were allowed to report sick, the returns for January 1915 show an average of about 4,500 a day, chiefly from pneumonia and blood-poisoning.

Wilson wrote that 'The water and mud increase and are getting horrible. The longer days will be very welcome when

1. O.H., 1915, I, 28.

they come, especially to officers; the men do not mind so much.' But he gave no reasons to support this distinction of taste.

On Christmas Day, 1914, there had been no firing, and in many sectors the troops had climbed out of their trenches and, meeting in No-Man's-Land, had talked and exchanged gifts. But such a development met with the strongest disapproval at G.H.Q.[1] and the officers responsible were punished. It did not happen again.

G.H.Q. seems to have been slow in realizing that the unfortunate tactical siting of the line was making an important contribution to the 'wastage'.[2] Finally, when it was seen that the line must be altered, there never seems to have been any thought of achieving this by making local withdrawals and inviting the enemy to step forward into the 'bad ground'. Instead, a variety of small, but extravagant, attacks were authorized with the intention of straightening the line and eliminating some of the more tiresome German enfilade buttresses that dominated it.

It is hardly surprising, in view of the conditions under which they were ordered, that these were uniformly unsuccessful, in spite of being pressed with the utmost gallantry. Sometimes, very rarely, the infantry, or such small proportion of them as had survived the passage of No-Man's-Land, managed to evict the Germans; but by nightfall they were almost spent, ammunition was low, they were under continuous fire from the German artillery. The reliefs, stumbling across a flarelit waste where the sappers slaved to dig some pretence of a communication trench in the mud, some meagre channel that would afford protection in daylight, were as often as not cut down by the machine-guns before

1. In *1914*, Sir John French wrote of '. . . individual unarmed men running from the German trenches across to ours holding Christmas trees above their heads. These overtures were in some places favourably received and fraternization of a limited kind took place during the day. It appeared that a little feasting went on and junior officers, N.C.O.s and men on either side conversed together in No-Man's-Land. When this was reported to me I issued immediate orders to prevent any recurrence of such conduct, and called the local commanders to strict account, which resulted in a good deal of trouble.'

2. O.H., 1915, I, 218.

they got there. It became customary to 'send a company to relieve a platoon—only a platoon's strength will arrive'.

. . . . .

In the warmth and comfort of the Allied Headquarters, however, the mood was one of optimism. Charteris expressed the general view when he wrote home: '. . . don't believe Captain M. that the war will last another two years. Germany has shot her bolt here and failed. . . .' It was more prophetic, if also provoking a more ominous reaction, when:

'. . . General Rice, our senior sapper, has made the most original forecast of all! He predicts that neither we nor the Germans will be able to break through a strongly defended and entrenched line, and that gradually the line will extend from the sea to Switzerland, and the war end in stalemate. *D.H. will not hear of it.* He thinks that we can push the Germans back to the frontier, and after that it will only be a matter of numbers.'[1]

Foch went even further, and thought that the time was already ripe. 'We are in a perfect condition, both morally and materially, for attacking,' he told Henry Wilson, whose diary also records:

'Long strategical talk [with Foch] in which we agreed that Germany still has one chance, and one only, namely, to shorten her front—and retire to the line Liège–Metz, or possibly even to the Rhine. Any middle course would be fatal to her.'[2]

In the closing weeks of the year preparations for a great winter offensive to accelerate this process were eagerly rushed forward; so eagerly, indeed, that a number of important considerations were overlooked, chief among these being the waterlogged state of the terrain and the dismal condition of the soldiers themselves. Then, at the last moment, Sir John French, who was to co-operate in the

1. Charteris, *G.H.Q.* The italics are mine.—A.C.
2. Wilson, *Memoirs*, p. 188.

north, lost his nerve and 'impressed on every commander that he was not on any account to get ahead of his neighbours in the attack; everybody was to wait for the man on his left'. And in the event everybody did wait, including the left-hand man. Thus the offensive proved 'not merely a failure, but a fiasco. The only effect produced was on Franco-British relations.'[1]

This abortive operation also had deep and significant psychological after-effects. In the first place the security of Sir John French's position was further undermined, both in his own estimation and in reality. Wilson was sent to Foch to plead against any complaint that might be sent out from G.Q.G.

'I made the best case I could about advancing in echelon from the left, and he listened without saying a word. At the end he said, "*Mais mon cher Wilson, nous sommes militaires pas avocats.*" That exactly expresses the straits I was pushed to. We discussed everything and he was as nice as could be; but "*Père Joffre n'est pas commode*", and it was clear that Sir John would be in a very difficult position if he did not put up some fight.'[2]

Wilson's mediation was of little use. Huguet noted that 'their [those of Joffre and French] relations which had never been trusting or cordial became colder and colder'. And there is no doubt that this made the Commander-in-Chief more jittery and indecisive than ever.

But more important, because more lasting, was the slur—as it was thought to be—left on the prowess of the Expeditionary Force. The French now openly declared that 'it might be helpful to hold the line and act defensively, but would be of little use in an attack'. Determination to redeem this and their own reputations was responsible for many of the worst excesses of stubborn leadership among the British commanders in the years to come.

1. Liddell Hart, *History of the World War, 1914–1918.*
2. Wilson, *Memoirs*, p. 192.

3

The First Experiment, at Neuve Chapelle

We are now about to attack with 48 battalions a locality held by three German battalions.

From a Special Order of the Day to the 1st Army, issued on 9th March 1915

AS THE bleak winter days dragged past there was little sign of improved relations between the Allied commanders. Joffre, who was accumulating troops for a further offensive effort in the spring, had asked French to relieve the 9th Corps, north of Ypres. The request was a reasonable one,[1] but the Field Marshal objected to its 'tone'.

'Sir John showed much anger at the tone of Joffre's letter, brought by Belin last night. And Sir John, who had arranged to meet Foch and Belin at Cassel at 11 a.m. this morning, refused to go.'[2]

Foch reduced the French demands to the relief of one division only, from the 9th Corps, but 'Sir John refuses to relieve anybody before April 1st'.

Joffre, who knew from Henry Wilson the rate at which the Expeditionary Force had been sent replacements, was furious. When Wilson went to Chantilly the following week Joffre began, at dinner, by loudly remarking: 'Well, your

1. As a glance at the map will show the French 9th Corps was isolated from the main mass of manœuvre of the French Armies by the stretch of line occupied by the Expeditionary Force. (The country to the north of Ypres was virtually impassable as a result of the inundations, and remained static throughout the war, being held by the remnants of the Belgian Army.)

2. Wilson, *Memoirs*, p. 208.

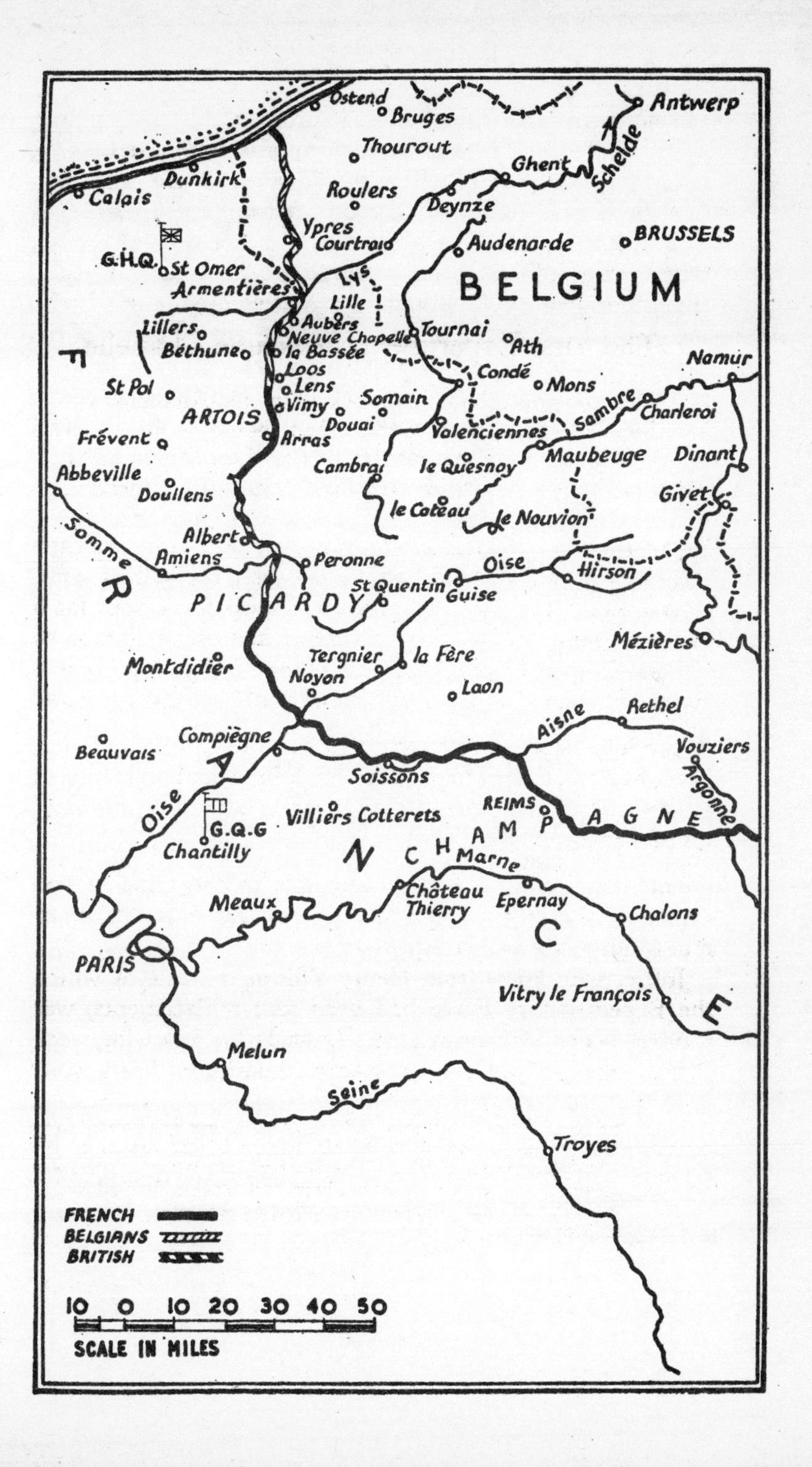
Ostend
Bruges
Antwerp
Thourout
Dunkirk
Calais
Roulers
Ghent
Scheldt
Deynze
Ypres
Courtrai
Audenarde
BRUSSELS
G.H.Q.
St Omer
Armentières
Lys
Lille
BELGIUM
Aubers
Neuve Chapelle
Tournai
Lillers
Béthune
la Bassée
Ath
F
Loos
Condé
Namur
St Pol
Lens
Mons
Vimy
Somain
ARTOIS
Douai
Sambre
Charleroi
Valenciennes
Frévent
Arras
Maubeuge
Dinant
Abbeville
Cambrai
le Quesnoy
Doullens
Givet
le Cateau
le Nouvion
Somme
Albert
Amiens
Peronne
Oise
Hirson
R
St Quentin
Guise
PICARDY
Mézières
Tergnier
la Fère
Montdidier
Noyon
Laon
Aisne
Rethel
Compiègne
Vouziers
Beauvais
A
Soissons
Argonne
Oise
REIMS
G.Q.G
Villiers Cotterets
CHAMPAGNE
N
Chantilly
Marne
Château Thierry
Meaux
Epernay
Chalons
C
PARIS
Vitry le François
E
Melun
Seine
Troyes
FRENCH
BELGIANS
BRITISH
10 0 10 20 30 40 50
SCALE IN MILES

Chief's extremely tiresome.'[1] As proof of his own power, Joffre related how he had been given 'absolute carte blanche till May' as a result of telling the Cabinet 'If you take away one single man that I can use on my front *I will resign.*'

The implication was plain. And in fact Sir John, also, was trying political pressure to secure his material needs, though applying it less blatantly. At this time it was by no means settled that the main mass of the armed forces of the Empire was to be fielded in northern France; other theatres seemed equally promising to sections of the Government in London. An expedition to the Near East, a massive naval effort to force the Kattegat—both these concepts had their supporters at home. Moreover, French must have felt doubtful as to how enthusiastic Kitchener would be over allowing the six 'new' armies, each of three corps, that would become available during the coming year, to swell the command of someone with whom he was on terms of almost open hostility.

In an attempt, as he said, 'to clear the air' and, at the same time, to emphasize his own importance, French wrote direct to Asquith, behind the back of the Secretary of State, 'a letter on the general situation' with particular emphasis on 'the manner of employing the New Army', and Asquith had it printed and circulated to members of the Cabinet.

Kitchener was furious, the more so as he had just completed plans with the Admiralty for an offensive against Ostend and Zeebrugge. But he was overruled in the Cabinet, the majority of whom seem to have felt, perhaps unjustly, that he was allowing strategic considerations to be subordinated to his personal distrust of Sir John French. Moreover, French had first taken the precaution of consulting his corps commanders[2] who had, not unnaturally, concurred, so that the whole of the Army in France was presenting a uniform opinion. It was necessary, though, for there to be

1. Wilson, *Memoirs*, p. 216.
2. Haig describes the summons to French's house in St. Omer in his entry for Monday, 4th January 1915.

some action that would justify priority for the French theatre and the command there, as well as restoring faith in British fighting ability at French General Headquarters. Haig had told the military correspondent of *The Times* that 'as soon as we were supplied with ample artillery ammunition of high explosive I thought we could walk through the German lines at several places'.[1] But both he and French realized that an early demonstration of this truth was necessary to establish their position. French wrote that 'Many vital considerations induced me to believe that a vigorous offensive movement by the forces under my command should be planned and carried out at the earliest possible moment.'

Unfortunately two adverse factors demanded consideration. In the first place, the ground was so sodden that any serious forward movement would be suicidal until the beginning of April, as attempts to dig in in fresh ground would strike water at a depth of eighteen inches. And in the second place, Joffre was now backing down on the promises which he had made earlier of a joint attack to the south, against the Vimy Ridge. His ostensible reason was 'shortage of troops' arising out of French's refusal to relieve the 9th Corps, although the state of the ground and his preparations for a 'really shattering' blow at the Germans in May probably influenced him more. At all events, Sir John had now to decide whether to secure French co-operation by carrying out the relief that Joffre had originally demanded and thinning out his own army, or to launch the attack in isolation.

For some days he discussed the problem with Haig. Some sort of a demonstration was plainly, indeed urgently, necessary. A simple relief of the French 9th Corps would hardly be noticed by the English papers and, indeed, by thus spreading out his troops Sir John would be postponing still further the day when he could launch his own offensive. Haig had a plan for a 'battering-ram' attack against the German position at Neuve Chapelle, to take place 'as soon

1. *Private Papers of Douglas Haig*, p. 84.

as possible', and independently of any action by the French. Of course, strategically it was a preposterous notion for the small British Army to launch an offensive without French support. But tactically there was a chance, by making full use of surprise and a local superiority of numbers, of breaking the German line and inflicting a sharp reverse. Sir John decided to adopt the plan. With evident satisfaction Haig noted (13th February 1915) that French

'. . . would prefer to take the offensive on my front rather than from Ypres because:

'(i) He wished me to carry out the operation as he could never be sure of getting satisfactory results from Smith-Dorrien,[1] and (ii) Because my troops were better.'

Haig had been toying with the plan, on and off, throughout the winter. In many respects the situation was full of promise. As regards numbers he was never again to enjoy such favourable circumstances for taking the offensive. The Germans had taken every available man, gun and shell from the Western Front preparatory to their great offensive against Russia in the late spring, and their contempt for the abilities of the British Army—in an offensive role at least—had led them to denude their front in that sector even more extravagantly than elsewhere.

Moreover the area selected for the attack was a point where the German line jutted westwards, forming a salient, about 2,000 yards round, between Port Arthur and the Moated Grange, with the ruins of Neuve Chapelle village immediately behind its centre. And here the local superiority of their own artillery, arising largely from shortage of ammunition on the British side, had fostered among the

1. Smith-Dorrien's decision to stand and fight at Le Cateau in August of 1914 had saved the Expeditionary Force, it is now generally agreed. But this had been taken in defiance of orders to go on withdrawing from French, and in spite of the fact that Haig's own corps had retreated at such a pace that it had completely lost touch with the enemy, besides exposing Smith-Dorrien's flank. The grudge which both commanders bore against Smith-Dorrien as a result of his showing up their pusillanimity in this way had much to do with circumstances of his dismissal at the height of '2nd Ypres'. See below, pp. 88 et seq.

Germans a feeling of security that was not warranted either by the lie of their position—it was exposed to converging fire from three sides—its ability to withstand serious bombardment, or their own numerical strength. For in this area there were only six companies amounting to some 1,400 men with twelve machine-guns between them. They were 'entrenched' behind a single line of sandbag breastwork built up shoulder high back and front on the water-logged ground, the water-line being only a foot or so below the surface. The wire—that was in later battles to consist of entanglements staked into the ground and as much as a hundred feet across—was here little more than two rows of *chevaux-de-frise*, portable trestle-like structures that two men could lift to one side.

Against this flimsy barrier Haig proposed to throw no fewer than forty-eight battalions, that is to say approximately 40,000 men, a numerical superiority of thirty-five to one. The two batteries of eighteen-pounders that had looked after the sector during the winter were augmented by a further sixty batteries together with forty 4·5 in. howitzers and eighty-two siege and heavy artillery pieces whose task was to silence the enemy counter-battery fire. Close behind the line the Cavalry Corps and the Indian Cavalry Corps were held in readiness to ride through the gap.

1st Corps Intelligence had accurately predicted the shape of the enemy resistance, as it had their reserves both local (800 infantry plus a cyclist and a machine-gun company four miles behind the line at Ligny-le-Grand) and general (a forecast of 4,000 additional rifles within twelve hours and up to 16,000 from corps and army reserves around Lille by the evening of the second day). In view of this it might have been felt that Haig's plan was excessively cautious: with such enormous superiority to attack on a frontage as narrow as that selected was far too conservative, particularly as Intelligence told him that he could count on a ratio of at least two to one as late as forty-eight hours afterwards. However, when Allenby, commander of the Cavalry Corps, suggested[1]

1. At a conference at Haig's H.Q. at Mérville 26th February 1915.

that the attack frontage be broadened so as to avoid the possibility of confusion in the bottleneck during the follow-through, he was sharply reminded of his 'unfamiliarity with commanding masses of infantry'.

. . . . .

Starting in the last days of February began the assembly of the assault troops. Soon the rear areas were '. . . absolutely choked with men, G-wagons, temporary bivouacs, long lines of horses standing patiently' and, more sinister, the S.A.D.S., the Supplementary Advanced Dressing Stations, spotlessly clean as yet, and empty.

Two Corps, the 4th (commanded by Lieutenant-General Sir H. Rawlinson, Bt.) and Indian (under Lieutenant-General Sir J. Willcocks, K.C.B., D.S.O., etc.), were squeezed side by side into the tip of the phalanx, each leading with one division; the 8th under Major-General Davies, and the Meerut under Lieutenant-General C. A. Anderson. Here already were the seeds of trouble, for both these units would be converging from opposite sides of the salient towards each other, trailing their long 'tail' and chain of command back to corps level and only finding unity of direction once again when their reports reached the army commander, Haig, at his headquarters at Mérville. Things might be all right if the day's progress went according to plan, but there was always the possibility that some unforeseen event would dislocate the machine, whose administrative duality threatened to double the confusion that Allenby had foreseen.

More immediately serious was a muddle over the artillery dispositions on the left flank. In spite of the profusion of artillery and the sparsity of targets, the destruction of the enemy breastwork along the last 400 yards of front on the extreme left had been entrusted to a brigade of six-inch howitzers—the 59th and 81st siege batteries—which it was discovered later had not yet arrived from England. This meant that there would be no heavy fire on the enemy positions there, as the eighteen-pounders were of little use in breaking up the breastwork, although they could cut the wire and keep the defenders' heads down while the actual bombardment was in progress. The absence of the two siege batteries was noticed and correspondence with the War Office took place on the subject[1] but, inconceivable though it seems, no alternative arrangements were made to cover the gap. The two missing batteries finally appeared on the evening of the 9th—the day before the assault—and had not sufficient time to build up gun platforms or lay their telephone lines to the forward observing officers, much less to range on to their allotted targets. For these reasons they played no part in the bombardment of the following day: an omission that was to have far-reaching consequences.

The eve of the battle was wet. Rain fell steadily, blown into occasional gusts of snow by a wind that was just above freezing temperature. Low clouds drifted across No-Man's-Land leaving long streaks of clinging mist that followed the course of the waterlogged communication trenches; the gun-barrels dripped with condensation and by nightfall the greatcoats of the waiting infantry were soaked through. But morale was at its highest. Many of the men, including the whole of the 25th Brigade,[2] were going into action for the first time. This was the first offensive operation of the war, and the majority believed that it would be the last.

And when, at 7.30 on the morning of the 10th, the bombardment broke, the effect was stupendous. This was the strongest concentration of guns per yard ever before assembled—one that was not to be equalled until the closing stages of Passchendaele, over two and a half years later. The

1. On the 26th February the War Office informed G.H.Q. that the batteries would embark on the 1st March. On the 2nd March they wired that embarkation had been postponed until the 5th. On the 3rd March G.H.Q. wired requesting an immediate embarkation of the batteries but they were not shipped until the 5th and did not arrive at Estaires until the morning of the 9th (O.H., 1915, I, 84 fn.).

2. Under Brigadier-General A. W. G. Lowry-Cole, consisting of 2nd Lincolnshire, 1st Royal Irish Rifles, 13th London, 2nd Berkshire and 2nd Rifle Brigade.

defenders were completely swallowed up in a storm of smoke, black and orange, from the high explosive, speckled with the hard white flash of the shrapnel; huge masses of earth, bodies, fragments of the enemy emplacements were tossed up and blown into the sky again and again, so that the battlefield became quite dark.

At five past eight the artillery lifted their fire from the German line and on to Neuve Chapelle village itself and the English infantry clambered out of their trenches and into No-Man's-Land.

The centre of the attack was led by the Berkshires and Lincolnshires. They found the enemy virtually neutralized by the bombardment; the wire had been shattered and such defenders as had survived were too dazed by the shellfire to offer any resistance. Suffering only the lightest casualties, they passed on over the crumpled remains of the enemy breastwork to their first objective—the German 'support' line. It had been expected that this might prove a tougher obstacle—it had only received ten minutes' concentrated fire as compared with the twenty allotted to the front line—and the orders to the leading formations were to 'consolidate here once the position has been captured by reversing the parados, etc.'. But on reaching this objective the leading British battalions found not only that it was not defended but that it had plainly not been occupied during the winter, for the sides were crumbling and the trench itself was full of water. However, in spite of the complete absence of opposition here, Colonel Feetham decided not to attempt the passage of this 'obstacle' but to halt his men just behind it. Colonel McAndrew of the Lincolns had been mortally wounded while leading his men across No-Man's-Land. He died a happy man for he had caused himself to be held up so that he could see his men enter the German trenches; but he was sadly missed for, without his leadership, the Lincolns, too, decided to halt without passing over the old German 'support' line.

However, within less than half an hour the second row of

the phalanx had caught up with the now stationary first row. These troops, the 2nd Rifle Brigade and 1st Royal Irish Rifles, passed right through the Berkshires and pressed on into the ruins of Neuve Chapelle village, which was empty of the enemy. They crossed the main street and advanced on into the open fields beyond towards the line of the 'Smith-Dorrien trench', the old British defensive position that had been built in front of Neuve Chapelle in the autumn of 1914. It was thought likely, at G.H.Q., that the enemy was occupying this position, or at least had it in such a state of readiness that he could have put reinforcements into it when the battle started. But this was not so; the trench, like the 'support' line, was disused. However, it had been scheduled for a half-hour bombardment, which started, to the moment, on the arrival there of the leading elements of the Rifle Brigade. These troops were thus compelled to withdraw some hundreds of yards and wait while the artillery methodically churned up the ground all round a position that the infantry on the spot could see to be unoccupied.

Immediately the bombardment stopped the Rifle Brigade resumed their advance with the Irish Rifles on their right, and were soon established along the line designated as their objective for the day—and this within one and a half hours of the start of the attack. Ahead of them stretched the flat green countryside, lightly scarred by the shellfire but seemingly lifeless. The line of the Layes Brook could be followed from the stunted willow bushes along its bank, broken only at one point by the group of shattered cottages that was to form, the next day, the 'Layes Bridge Redoubt'—at that time, like the dense undergrowth of the Bois du Biez to the south, unoccupied by the enemy.[1]

In view of the extraordinary opportunity that seemed to be presented, and the fact that the only enemy soldiers visible were occasional scattered groups that could be seen

1. The 56th and 57th Infantry Regiment (German) diaries show one section of the reserve Machine-Gun Company arriving at the Layes Bridge at 2.45 p.m. but the two Jäger Companies not getting into position in the Bois du Biez until 6 p.m.

running eastwards without rifles on the far side of the Mauquissart road, Lieutenant-Colonel Stephens at once sent a message back to Brigade H.Q. asking for permission to proceed with the advance. The sending of this message, and the time taken awaiting a reply, imposed a most irritating delay on the troops who, by now, had the momentum of the advance thoroughly in their blood. The minutes slipped by into quarters, then halves, then whole hours. From their left and also away to the south on their extreme right came the intermittent rumble of artillery fire. But ahead was emptiness, silence. The birds could be heard. After fetid months in winter dug-outs the air was clean and fresh. In little groups the men stood about chatting; some of them lit fires, others cleaned their rifles, still others dutifully improvised small shallow earthworks. Then, after 11 a.m., a body of men in khaki could be seen approaching from the direction of Neuve Chapelle. Were these, at last, reinforcements with which the spearhead could be pushed deeper? But no, it was simply a detachment of the 2nd Field Company R.E., with instructions 'to set to work constructing trenches and erecting wire'. On their heels came another message, from Divisional H.Q., to the effect that 'no further advance was practicable'.

This occasion is of interest historically, as on the 10th March 1915 was effected one of only three clean breaches in the German line throughout the war in the west.[1] The complete failure to exploit it in a vigorous and urgent fashion was due to concern—which may well be felt to have been unwarranted—at two checks which had been administered on the extreme flanks of the break-through.

. . . . .

The more serious of these had been in the north, at the left-hand corner of the enemy salient. Here there was a

1. The others being the first day at Loos, 24th September 1915, and the first day of the 'tank offensive' at Cambrai on 20th November 1917.

stretch of the enemy emplacement about 400 yards long, whose destruction had been allotted to the 59th and 81st siege batteries of 6 in. guns. For reasons explained above, these guns never fired a shot and the worst that the Germans suffered was a spattering of fire from the divisional eighteen-pounders. This gave rise to a lot of smoke and noise but left the enemy position virtually undamaged, serving as little else than a warning gong that called the Germans to man their fire-step.

This section was assaulted by the 2nd Middlesex under Colonel R. H. Hayes. They attacked in three successive waves, climbing out into a storm of point-blank fire, and some measure of their bravery may be taken from one sentence in the *Official History*: 'It was thought at first that the attack succeeded in reaching the German trenches as no one behind could see and not a man returned.' In fact every man, and there had been nearly a thousand, was killed.

A further result of the failure here was that the 2nd Scottish Rifles, who had been put in on the right of the Middlesex, were subjected to a vicious enfilade fire that swept diagonally across No-Man's-Land from the inviolate section of the enemy breastwork. This caused them heavy casualties, particularly among the officers, of whom 90 per cent, including Lieutenant-Colonel Bliss, the commander, were killed while attempting to rally the men and wheel the attack southwards and round the enemy flank. On this set-back being reported to the Brigadier he decided to put in more infantry and ordered the Devons and the West Yorkshires from Brigade Reserve to follow round in the footsteps of the Scottish Rifles. This decision is perhaps less inexcusable than it seems at first sight. For at that time it was impossible to 'call up' artillery fire direct from the front line. The request and accompanying report had to go back through Brigade to Divisional Headquarters for consideration there. Conscious as he was of the urgency of the situation—his headquarters were only 200 yards behind the front line—Brigadier-General Pinney felt that if extra infantry could do the job time would be saved. However, the renewed assault and the outflanking movement that was to accompany it took some time to get under way and, at the outset, suffered severely. Before they were properly developed, word came back from Divisional H.Q. that a further bombardment of the enemy position was to be put in hand at 11 a.m. and in the meantime the infantry were to be withdrawn to a safe distance. On this news being reported back to 4th Corps, immediate instructions were sent out to all the leading formations to halt and 'consolidate', and this, of course, included the unopposed battalions of the Rifle Brigade and the Royal Irish Rifles in the centre.

In holding up the whole of the offensive in their sector the 4th Corps were also influenced by the news from La Croix Marmuse (which was the H.Q. of the Indian Corps to the south). This told them of another check—less serious but leading to as great a muddle—on the extreme right of the breach. Here at zero hour the Gharwali Rifles had confused their direction on leaving the trenches and borne right-handed, running head-on into a section of the German defences that had been unprepared for assault by artillery bombardment. They had suffered heavily, losing all their British officers, but managed none the less to break into, and establish themselves across, the enemy position. In the centre the 2nd Leicestershire and 2/3rd Ghurka had a comparatively easy passage of No-Man's-Land but, owing to the loss of direction by the Gharwalis, there existed between these two forces a strip of enemy line, that had been on the fringe of the bombardment area and in which the defenders soon came to their senses and began to play fire up and down No-Man's-Land to their right and left.

The moment that news of this check filtered back to Divisional and thence to Corps Headquarters orders were sent out for all forward units of the assault to stop in their tracks and 'dig in', while steps were taken to eliminate the offending section. It was intended to mount an infantry attack with the troops most immediately available, the Seaforth Highlanders and the 3rd London, but, as will be seen in the following chapter, this took an inordinately long time to get under way. And while it was awaited all forward movement of the Indian Corps, as in the 4th Corps to the north, was halted.

All this time, although the movement of the leading formations had been stopped, the mass of the support battalions of both corps continued to press forward on their allotted timetable. Thus the effect of artificially retaining the cork in the neck of the bottle was an intense congestion in the rear areas. The slow traffic of the reinforcements travelled remorselessly up to the old front line, there petering out aimlessly among the shell-holes. Against it ran the first of the wounded; in their midst the engineers, cable layers, artillery observation officers and 'runners' attempted to carry out their tasks. All forty-eight battalions were on the move, but the fire-power of the phalanx was still limited to the strength of the troops deployed at its tip; the remainder were so much useless cannon-fodder and only the weakness of the enemy artillery at that time allowed what was, for all too many, an extra twenty-four hours of life.

4

Neuve Chapelle; the Passing Hours

. . . by 10 a.m. there were eleven battalions, roughly 10,000 men, in that narrow space. They lay, sat or stood uselessly in the mud, packed like salmon in the bridge pool at Galway, waiting patiently to go forward.

CAPTAIN G. C. WYNNE, FF XII 501

THREE hours after the assault, then, the position was as follows: on the extreme left of the line—that is to say, at the northern end of the attack frontage—the infantry had been disengaged from the enemy positions in the region of the Moated Grange and were awaiting a second bombardment before going forward.

To the south of this sector was a gap of about a quarter of a mile of flat country, traversed by a track, Signpost Lane, and the old enemy communication trench that ran alongside and bore the same name. On the far side, at a distance of about 200 yards, ran the sunken road that led straight into the rear of the line that was holding out against the remains of the Middlesex and the Scottish Rifles. The approach to this was unprotected other than by a few survivors of the enemy garrison who with one machine-gun were attempting to put the shattered group of cottages on the Mauquissart road into a state of readiness. But, unaccountably, there were no British troops in this area to take advantage of the gap—the nearest unit was the left flank of the Irish Rifles, to the east and south, on the line of the old 'Smith-Dorrien trench' and under instructions to 'dig in'. Further on down the line the Rifle Brigade, also completely unopposed at this

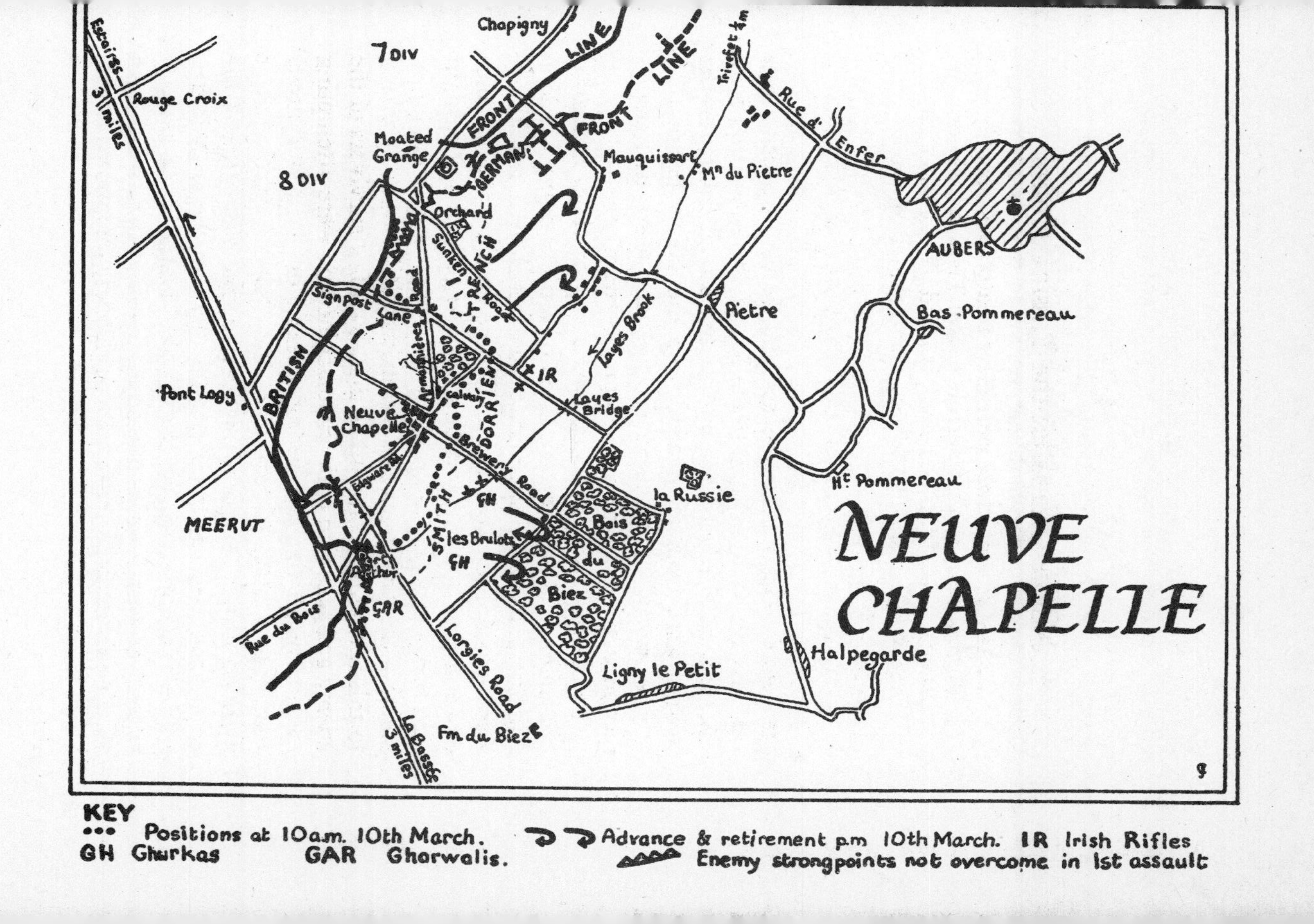

KEY
••• Positions at 10 a.m. 10th March.
GH Ghurkas
GAR Gharwalis.
Advance & retirement p.m 10th March.
IR Irish Rifles
Enemy strongpoints not overcome in 1st assault

time, was strung out with the ruins of Neuve Chapelle at its back. Then, exactly in the centre of the breach, was another gap, at the junction of the two corps on either side of Brewery Road, the lane that connected Neuve Chapelle with the Bois du Biez, and ran into the British front at right angles.

Below this line was the territory of the Indian Corps. Their attack frontage had been split into two by the deviation of the Gharwalis described in the preceding chapter. This had left an active section of the enemy line, again about 400 yards long, separating the right flank from the units in the centre. These units, the Leicesters and the Ghurkas, were ranged along the Layes Brook and the back of the 'Smith-Dorrien trench' up to a point just below the edge of Brewery Lane. Like the centre formations of the 4th Corps to the north, these troops were quite unopposed. Yet they adopted strictly defensive positions and made no effort to push forward scouting parties or to establish a forward screen. Had they done so, they would have found the Bois du Biez, a perfect natural strongpoint commanding the south-east approaches to Neuve Chapelle, to be unoccupied, and would have been able so to command the approaches as to prevent any reinforcement of the thin enemy front by the two Jäger companies stationed in Halpegarde.

But it seemed impossible for any further advance to take place, either as a result of direct orders (as in the case of the 4th Corps) or lack of initiative on the part of the commanders on the spot (as in the Indian Corps), until the flanks had been 'cleared'. No trace of urgency is detectable, however, in the direction of operations to achieve this purpose.

In the south, the task of clearing the Germans out of the Port Arthur salient had been entrusted to the Seaforth Highlanders, who were to attack down the trench line from the north while the 3rd London were to make another frontal attack. But the Seaforths were not in position until midday and then got so dispersed and confused in the course of their flanking march across the morning battlefield that the divisional commander was led to the assumption that they had already delivered the attack and had been held up. Accordingly a fresh barrage was put down on the northern sector of the offending position, arriving almost simultaneously—it was by now 2.15 p.m.—with the Seaforths, who had to withdraw immediately to get out of the way of their own artillery.

In the meantime the Germans at the northern end of the breach, in the region of the Moated Grange, had surrendered after a second bombardment and this flank was now completely cleared. However, General Rawlinson was convinced that the small orchard, which lay just behind the enemy position with one wall running along the side of the sunken road, was also a strongly defended strongpoint. There was no evidence that this was so, nor had Intelligence reported it, but General Rawlinson none the less instructed General Davies (commanding the 8th Division) that the men must be halted and a formal assault on the orchard prepared. A message from the commander who was on the spot (Brigadier-General Carter), to the effect that no sign of the enemy was detectable at any point in the orchard, took over an hour to travel back to Lestrem and crossed one from Rawlinson to the effect that the bombardment had been set for 12.30 and the assault was to go in immediately afterwards. In view of the fact that the whole of corps artillery was shortly to be put down on the orchard, Brigadier-General Carter was understandably reluctant to reconnoitre there and another half-hour's delay ensued. Finally, after the bombardment was over, the 'assaulting' troops moved in, and found the orchard neither defended nor even prepared for defence.[1]

At last there was no excuse for further hesitation by the left wing. It is true that the complete absence of enemy forces which had characterized the greater part of the front that morning no longer applied, for small groups of machine-gunners and other scratch units had filtered up to occupy

1. This episode should be borne in mind when reading the exchange of letters referred to on pp. 72–3.

the skeleton outpost line which the enemy had been constructing along the diagonal Mauquissart–Layes Bridge, but their numbers were very few—less than a hundred. On the other hand the British van had now become so swollen as to be highly unwieldy, even by a skilful commander with a clear idea of objectives and their priority. For the 'assault' on the orchard a number of units that had been accumulating just behind the line were pressed into action, so that the same length of front (i.e. that from Moated Grange to the left flank of the Irish Rifles on Signpost Lane) that had earlier been the province of the Middlesex and 3rd London was now, besides their remnants, crowded with the 2nd Wiltshire, the 2nd Green Howards, the 2nd Royal Scots Fusiliers, the 2nd Northamptonshire, the 1st Worcestershire and the 1st Sherwood Foresters.

In spite of, or because of, these massive numbers the advance was hesitant and dilatory in the extreme:

'It took a long time to get under way as the men were very thick on the ground and there was a good deal of sorting out to be done. We advanced in platoon columns of sections, preceded by a posse of men from other units bearing enormous light-coloured planks with which to cross the Layes Brook. These served to draw what fire was about.

'When we had progressed a few hundred yards against very mediocre opposition, word was passed along the line to halt the advance. I gathered that something had gone wrong. Actually all it was was that the lines of advance of two brigades had crossed. We lay in the open under desultory, but increasing, shellfire, for some hours. Ahead the Moulin de Pietre was clearly visible among the trees [two days later the Grenadiers were to be practically wiped out attacking it]—at that time it was unoccupied.'[1]

In the meantime Colonel Stephens had been desperately sending messages back from his position in the centre, urging the need to advance immediately. No reply was made to

1. Ewing, *The Royal Scots*.

these until 1.15 p.m. when the instructions read simply that the men were '*to be prepared* to advance towards Aubers ridge'. And at 1.30 Rawlinson sent a further message, this time to the mass of troops at the northern end of the breach who were fumbling forward after occupying the orchard, to the effect that they were to '*reassemble* as rapidly as possible *with a view* to a further advance'. The commanders on the spot can hardly be blamed for halting their men in perplexity at such instructions!

Immediately after sending these orders, Rawlinson sent a summary of the situation to Haig at Mérville. In this he stated that it was his intention to give the order to advance on Aubers at 2 p.m. Haig wired his approval, at the same time allotting still further troops to the 4th Corps from 1st Army reserve. But when, five minutes later, Rawlinson telephoned to La Croix Marmuse and spoke to General Willcocks he was told that the Indian Corps were still not ready to move forward as the Germans holding out in the Port Arthur salient had not yet been eliminated. Rawlinson therefore postponed, indefinitely, the issue of his orders to advance on Aubers. There followed an hour of complete inactivity. Then, at 2.45 p.m., came an enquiry from Mérville: General Haig wanted to know what was happening.

Rawlinson replied that he was waiting for the Indian Corps; but when Haig's chief of staff telephoned La Croix Marmuse Willcocks maintained that his position had been cleared up and that he was waiting for word from Rawlinson. The Meerut Division, he said, had already been ordered to attack the Bois du Biez. So once again the 4th Corps administrative machine creaked into motion. A message was sent back to the 1st Army saying that 'they were about to issue orders for the advance', and Haig ordered up yet more reserves—this time the 5th Cavalry Brigade. But the orders themselves[1] envisaged an advance of absurdly limited scope—when the second objectives designated were at a distance of less than a thousand yards.

1. O.H., 1915, I, Appendix 17.

Whether on account of a leisurely atmosphere generated by such unambitious instructions, or on account of the heightening congestion both behind and in the front line, it was a considerable time before these orders were put into execution. They were received at Divisional Headquarters just after 3 p.m. but had not percolated to the various brigades concerned until just before four o'clock. The attack had, in fact, been ordered for 3.30 and so was already out of phase with the artillery bombardment. The gunfire was rendered still less effective by the fact that none of the new targets could be clearly seen from the existing observation posts, nor had they been previously registered. The only effect of the haphazard and sporadic fall of shell that followed was to inflict casualties on some of the Devons who had pushed ahead of the line and on a group of ruined miners' cottages along the Mauquissart road. The Devons withdrew, and about half an hour later the Germans re-occupied these cottages and placed two machine-gun sections in the cellars. (They were never evicted, either on that day or those following, and used this point as one of the pivots of their northern line.) Further delays were caused by an absurdly protracted exchange of messages between Brigadier-General Carter and Brigadier-General Watts, each enquiring whether the other was ready, their messages crossing in transit and leading to fresh and differently worded replies, and so forth. When, finally, the leading battalions moved forward across the Armentières road and in the direction of the enemy it was after 5.30 p.m.

By this time thick clouds that had been gathering during the afternoon had formed an unbroken ceiling over the sodden battle area. A grim twilight shrouded the unfamiliar terrain, masking the little clumps of trees, the shattered farm buildings, the stunted willows that marked the course of innumerable swollen brooks. And all the while the German strength had been increasing so that as the British infantry groped their way forward in the dusk there could be heard once again, in increasing volume, that most haunting of all the sounds of trench warfare—the drawn-out clatter of a long burst on the machine-gun.

. . . . .

Down on the Indian Corps front the renewed forward movement ordered by General Willcocks had got under way earlier, though still over half an hour later than ordered. And, as dusk fell and the damp mist rose up from the dykes and water-meadows, the leading Ghurkas reached the western edge of the Bois du Biez. At the southern end of the wood the 2/2nd Ghurka occupied the group of cottages, still burning from the morning's bombardment, known as 'les Brulots', and sent scouts on into the wood; but at the northern end the 1/9th came under intermittent fire from a German machine-gun in the cellar of one of the houses by the Layes Bridge and Lieutenant-Colonel Widdicombe kept back the greater part of the battalion behind the line of the Layes Brook until the 4th Corps should have cleared his left flank. (Ironic to think of these men hesitating at one machine-gun at that range when, less than eight weeks later, they were to be sent in repeatedly against a volume of fire twenty or thirty times stronger.)

In the meantime the meagre German reserves were taking advantage of the failing light to enter the wood, undetected from the eastern side. So it was that at dusk the Bois gradually began to fill with stealthy little groups of infantry, stumbling and crackling in the unfamiliar undergrowth. The air was quiet. The two armies were still poised, like boxers in the first round, nervous at the lightest feint. The artillery was silent as the guns trundled to their new positions; only to the north could be heard the intermittent chatter of the German machine-gun at Layes Bridge. As darkness deepened some rifle fire could be heard, and shouted orders, Hindustani mingling strangely with English and German.

Then, just after 8 p.m., occurred one of those tricks of

chance whose effect was to be amplified by the timid and hesitant leadership of the attackers. Ghurka scouts captured a corporal of the 56th Infantry Regiment, who stated under interrogation that two regiments (in reality only two battalions, but from different regiments) were collecting in the wood. When this was reported to Brigade Headquarters, they replied with instructions that all forces were to be withdrawn behind the Layes Brook for the night, and with them went any chance of reaping even a local success from the Neuve Chapelle offensive.

During the hours of darkness the enemy worked with prodigious energy to improve his position: the newly arrived infantry dug a line of shallow breastwork to connect the machine-gun nests at Mauquissart and Layes Bridge and these two strongpoints were further improved and armed with additional machine-guns. The Germans also ran an outpost line in front of the Bois du Biez, with their main body of machine-gunners lying in the fringe of the trees so that here there was defence in depth.

Haig's orders came through at dawn on the 11th. They ordered an attack at 7 a.m. '. . . to begin at all points, and to be pressed vigorously, as from information received it appears that the enemy before us is in no great strength'. Although this estimate of the enemy strength was still true, relatively speaking, the task of the 1st Army was now immeasurably more difficult. The element of surprise had gone, the infantry planning had been thrown out of mesh, and the weight of artillery support was now very much less.

Fifteen minutes' bombardment was due to start at 6.45, but dull and misty conditions impeded observation and registration of targets. Many of the guns never ranged properly on to their targets and the new enemy breastwork, being as yet undetected, was not shelled at all. When the assault did get under way it followed the same principle of 'congested development' as on the previous day, with brigades uncovering each other and diverging as they moved forward. But now the confusion latent in such practice was aggravated by the much-increased enemy resistance. The Germans had brought up several fresh batteries in the night, and their bombardment began at 6.30 and lasted for over three hours, deluging Neuve Chapelle village and the whole rear area with shellfire and gravely impeding movement and communication. The 20th Brigade, which had come up fresh, the Grenadier Guards and Gordon Highlanders leading, was badly knocked about as it passed through the 21st Brigade lines opposite Mauquissart and soon got lost among the maze of ditches that intersected the area. After edging forward through the smoke and under fire the whole time they halted some forty minutes later astride a deep drainage dyke mistakenly supposed to be the Layes Brook. When their position was reported back to Divisional H.Q. it was taken to mean that a complete break-through of the German lines had been effected and the artillery was lifted. This had the effect of making any further advance impossible, yet owing to the telephone lines having been cut by the enemy bombardment the higher commanders remained in ignorance of the state of affairs in the firing line until the early afternoon.

Another factor contributing to the muddle was that, for some reason which has never been explained, there had been no relief of the 2nd Rifle Brigade, the force exactly in the centre of the wedge. They had, indeed, received no further orders since their instructions to 'consolidate' at 12.45 p.m. the previous day, and now, instead of finding themselves opposite a vacuum, they were directly facing the strongest part of the enemy line—the 'Layes Bridge Redoubt'—which had been reinforced with an additional twenty-two machine-guns during the night of 10th March and the morning of the 11th.[1] Colonel Stephens, their commander, who had had his urgent pleas for permission to advance rejected or ignored the previous day, was now naturally reluctant to make any further movement forward as his men were short of food and ammunition as

1. Regimental diary, III Jäger Battalion.

well as exhausted physically. But the weakness here did have the serious tactical effect of further emphasizing the division of the front between the 4th and the Indian Corps, the one to the north, the other to the south of the Layes Bridge, and each functioning only along the ponderous chain of command that led back to their respective Corps Headquarters.

Whether this failure to relieve 2nd Rifle Brigade arose from oversight, or from General Rawlinson's preference for independent action in a north-easterly direction, is not clear. The first mention of an intended relief is at 10.25 a.m. on the morning of the 11th, but General Rawlinson was convinced that a German counter-attack would follow on his own failure to breach the Mauquissart road position that morning, and cancelled the relief order at the last moment, directing the reserves to support along the axis of the Pietre road—an area already desperately overcrowded and carrying a heavy traffic of wounded.

In fact no German counter-attack did materialize in the forenoon—the enemy were still engaged in improving their defences and were certainly not sufficiently numerous to attempt an advance across the open. By 12.19 p.m. Rawlinson seems to have realized this for he sent out orders for a fresh attack, and that 'the objectives should be captured without further delay'.

However, the German artillery fire was increasing in severity the whole time, and all telephone communication with the front line had been cut. As a systematic diffusion of the new orders by runner would have meant delaying the attack for another two or three hours the unorthodox course was adopted of sending forward the support battalions (1st Worcester and Sherwood Foresters) with instructions to 'go straight in to the attack and carry the front-line troops forward with them'. They set off late, five minutes, in fact, after the supporting artillery fire had stopped, and suffered heavily in crossing the open ground that led up to the foremost positions. On arrival they met with a flat refusal on the part of the officer in command to go into the attack a second time:

'I received a note from the Worcestershire: "We have *got* to advance, will you give the order?" I answered: "No, it is a mere waste of life, impossible to get 20 yards, much less 200. The trenches have not been touched by the artillery. If artillery cannot touch them the only way is to advance from the right flank. A frontal attack will not get near them." '[1]

In spite of this the two leading companies of the Worcestershire were ordered over the top by their own commander; here they were shot down almost at once. On the left the Northamptons made three attempts and some of their number managed to reach some dilapidated farm buildings in No-Man's-Land where, in due course, the majority of them were killed by enemy artillery fire. The fate of Colonel Prichard, after his courageous refusal to subject his men to further pointless slaughter, is not recorded.

The position by nightfall on the 11th, then, was simply that no further progress had been made but that casualties were rising hourly. Simple mathematics should have forced the decision to break off the engagement—the British artillery was weaker and less accurate than on the day of the assault, the infantry were exhausted and in some confusion while the enemy was nearly five times as numerous as on the previous day. However, General Haig was determined to proceed with his plan. On the evening of the 11th he visited the support area and '. . . gave personal orders for guns to be brought closer to the front in several places'. On return to his headquarters he issued orders for a 'simultaneous advance' by the 4th and the Indian Corps[2] at 10.30 a.m. the following day after a half-hour artillery bombardment—an almost verbatim repetition of his orders of the morning before.

1. Message received at 24th Brigade H.Q., 2.50 p.m., 11th March 1915, from Colonel Prichard.
2. O.H., 1915, I, Appendix 20.

On this day, the 12th, 1st Army was given a last chance —by the stupidity of their opponents. For at dawn the Germans had launched a counter-attack with sixteen battalions along the whole length of the line. This was easily repulsed, but no effort was made to follow up their confusion and the precious morning hours were allowed to slip by until the zero fixed on the previous day. Even this seemed too early, for at 9.20 a.m. Rawlinson telephoned to Haig[1] and told him that 'unsatisfactory' artillery registration threatened to jeopardize the effectiveness of the bombardment and asked for zero to be put back two hours. Haig consented to this, but when, at 11.15, he was again told that the gunners were not ready he directed that 'forward movement should not be postponed any longer'.

At 12.30, accordingly, the assault went in. Within two hours it had been brought to a dead stop, with gains measured quite literally in yards and with many of the leading units virtually annihilated. But there were still, concentrated in the immediate neighbourhood of Neuve Chapelle village and the old German front line, large numbers of troops who, if not fresh, had not, at least, seen direct action, and Haig gave orders that these should now be committed. At 3.6 p.m. he wired all units that

'Information indicates that enemy on our front are much demoralized. Indian Corps and the 4th Corps will push through the barrage of fire regardless of loss, using reserves if required.'

It is very hard to decide what was in Haig's mind at this time. Certainly the reports coming in to 1st Army gave no indication of any other result than 'loss' of life. Along the whole front a new pattern was emerging, of fresh units moving forward to the line under fire, intermingling and hesitating briefly among the battalions already shattered in the morning's attacks and then, themselves, attacking piecemeal with the inevitable result. In spite of this patent

1. O.H., 1915, I, 139.

collapse of the offensive momentum Haig seems to have thought that the answer was simply to throw in yet more troops. That afternoon he telephoned to Sir John French asking that the 2nd Cavalry Division should be placed at his disposal, which request was granted, and even as late as 6.20 p.m. there is a record of a telephone conversation with G.H.Q. in which Haig asked that

'In view of the promising situation, the 46th Division from General Reserve might be used to relieve the two left Canadian Brigades . . . and that these should be massed with a view to breaking through opposite Rouges Bancs and co-operating with the advance of the 7th Division.'

However, those nearer to the front were less optimistic. Both Rawlinson and Willcocks passed on the orders to attack 'regardless of loss' and to 'push forward at all costs', but as these instructions reached brigade level they began to meet with a similar reaction to that of Colonel Prichard the previous day—although it found a more discreet expression.

In the Indian Corps the commander of the Ferozepore Brigade, Brigadier-General Egerton, ordered up from reserve with instructions to take over command of the Sirhind and Jullundur Brigades also, and to mount an immediate attack against the Bois du Biez, first succeeded in postponing the time of the attack to 8.30 p.m., then, after consultation with the front-line commanders, telephoned a request for a further postponement until 10.30 p.m., 'at the same time giving his opinion that the attack ordered was not likely to succeed'.[1] Thereupon General Willcocks cancelled the attack altogether, giving as his reason to Haig, who had turned up at Indian H.Q. at La Croix Marmuse a few minutes earlier to urge the corps to greater effort, '. . . that he did not consider it feasible to make an attack with such a large body of troops by night over unreconnoitred ground'. The cancellation of the Indian Corps attack must

1. O.H. 1915, I, 144.

by itself have sealed the fate of any offensive by 4th Corps to the north but the first orders cancelling their own attack did not start being received at the various brigade headquarters until 1.25 a.m., and by that time a number of ill-co-ordinated forward movements had already got under way. The confusion was very great and heightened by the exhaustion of the men who, after three days and nights continuously under fire, had fallen asleep and 'could only be aroused by the use of force—a process made very lengthy by the fact that the battlefield was covered with British and German dead, who, in the dark, were indistinguishable from the sleepers'.[1]

In fact the offensive had run down completely, and by the morning this was plain, even at 1st Army H.Q. Haig therefore issued orders that the leading units were to consolidate their present positions and those in close support were to be taken back into reserve, with the intention of pushing in a second offensive within the next few days with two fresh divisions.

Happily, this scheme did not come to fruition and, as the days passed, it came to be realized that 1st Army had, although 'gaining valuable experience', suffered a defeat.

As to the causes of the defeat, opinion was less than unanimous. Haig's diary for 16th March tells how

'Sir H. Rawlinson came to see me about 8 a.m. and handed me a letter which he had written about Major-General Davies commanding 8th Division. He did not consider him a good commander of a division on the field of battle. In forwarding on the letter to G.H.Q. I concurred in R.'s opinion, said he [Davies] had done well in preparing the attack, but that he had failed to advance from the village of Neuve Chapelle at once after its capture. I thought that he was unfit to command a division at this critical period of the operations in France but should be employed at home.'

1. O.H., 1915, I, 149.

But something seems to have gone wrong with this attempt to make Davies the scapegoat because later:

'. . . I received a letter from Rawlinson enclosing one from Davies. As a result of this, R. at once wrote that he took all responsibility for delaying the advance from the village until 3.30 p.m. This at once showed that Rawlinson felt himself to blame for the delay, not Davies. I am afraid that Rawlinson is unsatisfactory in this respect—loyalty to his subordinates, but he has many other valuable qualities. . . .'

The background to this episode is not clear.[1]

If relations between the British generals had temporarily deteriorated, those between the Expeditionary Force and the French showed a marked improvement—and for an ominous reason.

'After the battle of Neuve Chapelle the correspondence between the French and British general staffs contained a fresh note of confidence and a more cordial desire on the French side for an effective co-operation in a combined offensive movement.'[2]

For more important than any other 'lesson' was the fact that the British commanders had shown their readiness to attack 'regardless of loss' even if loss was to be the only result.

Prophetically, Charteris wrote:

'I am afraid that England will have to accustom herself to far greater losses than those of Neuve Chapelle before we finally crush the German Army.'[3]

1. See p. 61 above, and footnote.
2. O.H., 1915, I, 154.
3. Charteris, *G.H.Q.*, 86.

5

Gas

> The horrible part of it is the slow lingering death of those who are gassed. I saw some hundred poor fellows laid out in the open, in the forecourt of a church, to give them all the air they could get, slowly *drowning* with water in their lungs—a most horrible sight, and the doctors quite powerless.
>
> CHARTERIS, 28th April 1915

WITH the coming of April the weather had improved. One glorious spring day followed another, drying out the scabs of war that lay upon the soil. On the 22nd temperatures were in the seventies.

Throughout that morning the Germans had been putting heavy shell, 8 in. and 17 in. howitzers, on Ypres and the roads leading out of the town, but during the afternoon all was quiet. The sun shone gently on the flat countryside, glinting on the *Beekes*, the deep drainage ditches that meandered across the fields, still swollen with the rains of winter. Everywhere the green shoots of spring struggled upwards through the filth and squalor of battle. Right up to the support lines the land was still under cultivation. For four months the salient had seen no serious fighting.

And then at 5 p.m. a new and furious bombardment of the town and the villages in front and to the north of it began. The enemy artillery, that had ranged accurately on to all these targets in the morning, now searched them out in crippling strength, making movement almost impossible. Soon those observers who were on points of vantage saw

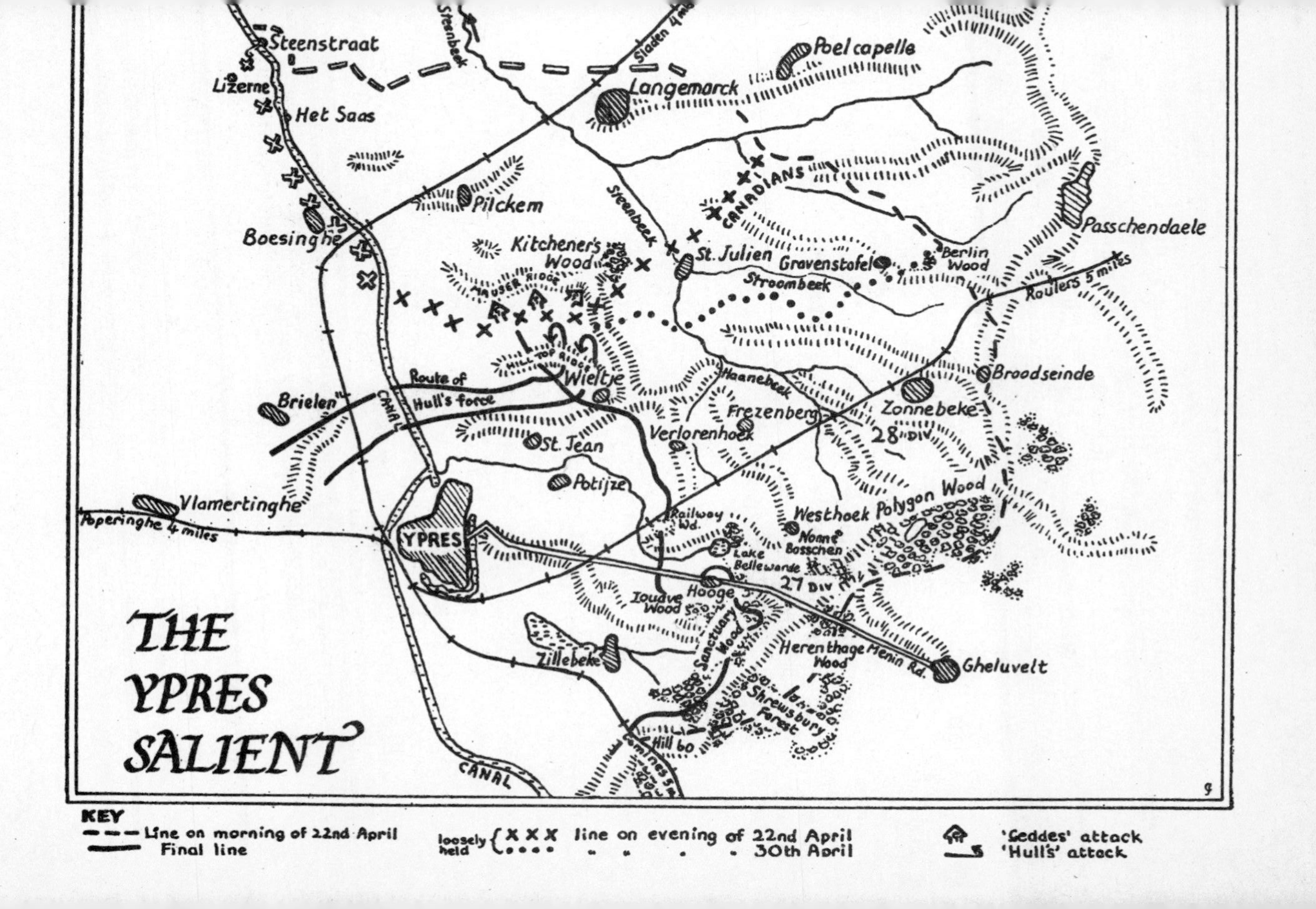
THE YPRES SALIENT
Steenstraat
Lizerne
Het Saas
Boesinghe
Pilckem
Langemarck
Poelcapelle
Passchendaele
Steenbeek
Sladen 4 miles
CANADIANS
Kitchener's Wood
Mauser Ridge
Hill Top Ridge
St. Julien
Gravenstafel
Berlin Wood
Stroombeek
Roulers 5 miles
Broodseinde
Zonnebeke
28th Div
Haanebeek
Frezenberg
Verlorenhoek
Wieltje
Route of Hull's force
Brielen
CANAL
St. Jean
Potijze
Vlamertinghe
Poperinghe 4 miles
YPRES
Railway Wd.
Westhoek
Polygon Wood
Nonne Bosschen
Lake Bellewarde
27 Div
Hooge
Zouave Wood
Sanctuary Wood
Herenthage Wood
Menin Rd.
Gheluvelt
Zillebeke
Shrewsbury Forest
Hill 60
Comines 5 m
9
KEY
Line on morning of 22nd April
Final line
loosely held
X X X line on evening of 22nd April
• • • • " " " 30th April
'Geddes' attack
'Hull's' attack

two greenish-yellow clouds creeping out across No-Man's-Land in the French sector, on either side of Langemarck. These clouds spread laterally, joined up and, moving before a light wind, became 'a bluish-white mist, such as is seen over water-meadows on a frosty night'. Rapid fire from the French 75 mm. field batteries could be heard, as could the rifle fire of the Germans who seemed to be advancing.

The Canadian Division, on the right of the French, had only lately arrived and had as yet no proper communication with them, while the telephone lines to and from Corps Headquarters had all been cut by the bombardment. Thus it was impossible for anyone in the 2nd Army to form a clear picture of what was happening, but, as the minutes slipped past, people in the rear areas, and in particular the British reserves in billets around St. Jean and La Brique which were due south of the French positions, became aware of a peculiar smell and smarting of the eyes. Then, as the enemy artillery fire lifted, dense masses of fugitives came stumbling down the roads from the direction of Langemarck and Pilckem. Few of them could speak, none intelligibly, many were blue in the face, others collapsed choking by the side of the road. At the same time onlookers began to feel a tingling of the nose and throat, and a tightening of the chest. At this time the mob was composed about equally of Tirailleurs, civilians and French African troops, but it soon became thicker and more disordered as the first of the French artillery teams and wagons attempted to drive their way through those on foot.

The air was heavy with fear, with the stark panic of the unknown. The few British units in the path of the retreat, the 2nd Middlesex, 3rd Buffs, and two and a half companies of Canadian troops, began—with a coolness and discipline that seems almost incredible as we look back on it—to prepare their positions for defence. It was obvious that something very serious had happened, although for about an hour the 'seventy-fives' of French divisional artillery could still be heard. Then, at about 7 p.m., these, too, suddenly and ominously ceased fire.

. . . . .

Shortage of troops had restricted the weight of the German attack, but their hopes that surprise and the introduction of the new weapon would foster its chances were more than fulfilled. For this area held by the French was weakly garrisoned. It had seemed that the deadlock was unbreakable here, and vigilance had relaxed. The trench-line faced north-east, running across very gently rolling country, from the Belgian positions along the Ypres–Dixmude Canal, in front of the town of Langemarck, up to the junction with the 4th Corps of the British 2nd Army on the Poelcapelle–St. Julien road. A distance of four and a half miles, it was covered by two French divisions, the 87th Territorial and the 45th Algerian. They were generously backed by artillery and had the added advantage of enfilade fire from the Belgian positions that were almost at right angles to their own. But the quality of the troops was poor. An officer of the Leinsters, holding the line adjacent, noted:

'These French native troops are a picturesque lot, in their blue tunics and baggy red trousers, but they are a confounded nuisance to us. They keep up a rapid fire all night, so that the enemy retaliates and often catches our working parties in No-Man's-Land; then, in the daytime, they drift back to loot in the ruins of Ypres. They never bury their dead and the whole place stinks to heaven.'[1]

This was in fact the old 9th Corps area, that had proved such a source of trouble between Joffre and Sir John French earlier in the year. It is not impossible that G.Q.G. were deliberately using it as a 'waste-paper basket' for second-rate units in the hope that trouble here would force the British into a *de facto* relief so as to safeguard the left flank of their own divisions in the salient, although the French cannot have

1. F. C. Hitchcock, *Stand To.*

expected that the 'trouble' would have been so sudden or so drastic.

Nonetheless, it must be recorded that both the French and the British had ample warning of the Germans' intention to use gas. On the 20th March some prisoners had been captured on the south side of the salient and under interrogation had given extensive details of the plan and of the placing of the cylinders in the trenches. The idea was regarded as being so fantastic, though, that it took some time to filter up the chain of command, and in the meantime the division was posted to a new area. When, finally, it reached Army Headquarters, it was duly published in the bulletin; but this circulated only in the Artois district, over 100 miles away. Then, a week before the attack, a deserter surrendered in the exact area that was to be attacked, near Langemarck, and supported his evidence by showing one of the crude respirators with which the German infantry there had already been issued. The French divisional commander was gravely impressed, but the corps commander, Balfourier, dismissed the concept as 'absurd' and administered a sharp rebuke at the manner in which the usual channels had been bypassed to warn the British and French units on either side.[1] Then, three days later, the Belgians captured further evidence of the enemy design, and again the information was not taken seriously above brigade level.

All these warnings might just as well have never been given for the heed that was paid them. The higher the rank the more ludicrous did the idea seem, the British commanders taking their line from Haig who, a fortnight previously, had given short shrift to a visitor whose mind was working on these lines:

'Lord Dundonald arrived from England. He is studying the conditions of War in the hopes of being able to apply to modern conditions an invention of his great-grandfather for

1. Advance warning of the gas attack, received at French and Belgian H.Q., is given in greater detail in Appendix 2.

driving a garrison out of a fort by using sulphur fumes. I asked him how he arranged to have a favourable wind!'

And now the front, from the canal to the Canadian flank, was broken, completely shattered. As dusk fell, the sound of rifle fire died away from the battlefield as the last pockets of resistance collapsed. The field-guns, too, were silent; those of the French had been overrun and the German infantry had long since outstripped their own.

'Where was the firing line? Nobody seemed to know. All around us there was a curious silence, but in the background you could hear the pounding of the "Jack Johnsons" [German 5·9 in.] on the town and the roads out. It was impossible to tell friend from foe and the place was still filled with little bands of French native troops. They were without officers and completely disorganized, some of them were in a very bad way and coughing up quantities of blood and pus. . . .'

Had the Germans realized the probable scope of their success there seems every likelihood that they could have pressed right through to the Menin Gate and taken the whole of the British force in the salient in a noose. For by the time that darkness had enveloped the battlefield there were no French troops left fighting east of the canal and the eight-thousand-yard gap was covered by a mere ten British and Canadian battalions, many of them under strength, without co-ordination, leadership, proper communication or prepared positions of any kind.

However, the enemy plan, limited by the absence of reserves, was for no more than an effective salient cut, that would disrupt the Allied plans for a spring offensive and, at best, compel the evacuation of the Ypres salient. To this end the German infantry, after advancing to a depth of about two and a half miles, halted at a shallow though significant rise in the ground, known as Mauser Ridge, that ran parallel with the old front. Here they began to dig in, placing their main line on the reverse slope and running a string of outposts on the forward side, with the flank resting in 'Kitchener's Wood', a dense copse that had, up until that morning, concealed a number of French and Canadian field batteries.

The British were strongest on the extreme right of the breach where the Canadians were still occupying their original positions. Here, although their flank was 'in the air', the Canadians had sufficient troops in divisional reserve to improvise a throwback line along the Poelcapelle–St. Julien road during the night, although the narrow triangle formed thereby must obviously prove untenable over a long period. However, at about 8 p.m. a liaison officer arrived at the headquarters of the Canadian Division with the news that the French 45th Division was going to counter-attack during the night against the other side of the gap and pleading for 'urgent and immediate' support.

In fact it was quite out of the question for any offensive operation to be mounted by the French 45th Division, which had virtually been destroyed as a formation and had lost all its artillery. However, this fact could not, on the scanty information available at the time, be appreciated at Canadian H.Q., and it was felt that the enemy might be compelled to withdraw by concerted pressure on the two sides of the breach. The 3rd Canadian Brigade was accordingly ordered to counter-attack against 'Kitchener's Wood' 'and then press on east of Pilckem' as soon as possible.

But although these orders were issued at 9.40, it was not until past midnight that the reserves were in position. Darkness and the unfamiliarity of the troops with the terrain and their task hampered the operation from the outset. Nor was the exact locality of the Germans known, so that the Canadians, when they finally got under way, had to advance virtually 'blind' towards the dark mass of 'Kitchener's Wood' some half-mile distant in the failing moonlight. At 300 yards the enemy opened fire and the Canadians broke into a run, their bayonets glinting in the green light of the phosphorus flares. Within minutes, although nearly half their number had been killed or wounded, they had fought their way deep into the wood and the enemy was in flight. By 2 a.m. they had reached its centre where stood the four sad guns of the 2nd London heavy battery, their long barrels still pointing impotently at the now retreating Germans.

But this success, bought at a high price in lives, was both local and short-lived. The French on the left flank had not moved, nor made any effort to distract the enemy. In the centre the British troops were still in too confused a state to have had any prospects of a successful night attack and, likewise, had made no demonstration. The Germans in 'Kitchener's Wood' had been little more than an outpost line and, on their withdrawal being signalled, the 5·9s began systematically 'squaring-over' the whole region with a hard insistent bombardment that started at dawn and continued with mounting intensity throughout the following morning.

With the exception of this gallant but ill-conceived sally the Allies had made no direct move against the enemy during the night, although there was so much confused marching and counter-marching as a result of contradictory orders that the majority of the men were thoroughly exhausted by daybreak. These conflicting instructions were issued by a variety of officers, each of whom supposed himself to be in complete charge, as the motley collection of units in the 'line' were each independent, having no proper command structure, or only one that ran back to some brigade and divisional headquarters outside the threatened area. For example, the Buffs and the Middlesex had been resting in reserve for the 28th Division, itself in position in the extreme eastern tip of the salient. At the very moment (one o'clock on the morning of the 23rd) that they received instructions to move north, the Germans launched a fierce local attack on the 28th Division front and on the neighbouring 27th Division that faced south. Orders were immediately sent out to the reserves of both these divisions to 'hold themselves in readiness to move up to the [south-eastern] line', and they had to turn about—no easy task in the darkness, when

F

'... every civilian left in the salient seemed bent on getting out as soon as possible, usually pushing a handcart and driving his four best beasts'.

Telephone communication was almost impossible as the enemy artillery fire had severed all the direct lines, and the liaison officers had to gallop their horses across the darkened, torn-up country, the main roads being impassable under the constant shellfire. None the less, by about 3 a.m. a motley collection of about 4,000 men[1] had been assembled, under the nominal command of Colonel A. D. Geddes of the Buffs, with orders to make a counter-attack 'as soon after dawn as was practicable'; and by 5.30 a.m. these men—the majority of whom had not eaten since midday of the 22nd—were deployed in positions for advancing. The extreme short notice at which these dispositions had to be made, allied to the other factors making for confusion related above, and the commanders' ignorance of the exact position of the enemy, had unfortunate consequences.

The British infantry were halted and strung out in the long lines of their attack formation at a considerable distance from, and out of sight of, the enemy. The Germans were dug in along Mauser Ridge. Running parallel with this, and to the south of it, was another gentle rise—Hill Top Ridge. The attacking infantry were drawn up short of the crest and were then sent off at irregular intervals 'in the general direction of the enemy', and without any pretence of serious artillery support. The moment that they breasted Hill Top Ridge they came under heavy fire which played over them all the time that they were descending the reverse slope and crossing the declivity between the two ridges (that was christened 'Colne Valley' later in the war). Here their condition was worsened by fierce enfilade fire from machine-guns dug in during the night on a spur of the Boesinghe Ridge that the French had mistakenly reported as still being in their own possession. Thus the attack never succeeded in coming to

1. Consisting of battalions of 2nd Buffs, 3rd Middlesex, 5th King's Own, and 1st York & Lancaster.

grips with the Germans, although the survivors managed to crawl up to within a few hundred feet of the enemy, where they scraped and improvised what cover they could, remaining pinned down, without water or other rations, until nightfall.

. . . . .

While this attack was in progress Sir John French drove over to see Foch at Cassel. Foch, as usual, was grossly optimistic. Crises had a stimulating effect on the French general and seemed to generate in him a kind of excitable euphoria. Reminiscing after the war he said:

'One knew nothing, one could know nothing, and if one waited till the next day it meant a break-through. I sent Desticker to Elverdinghe. He "legged" it all night long. During this time Weygand and I at Cassel were warning the divisions at Arras—they arrived at the rate of one a day—the gap was closed!'[1]

But, as his biographer drily comments, 'These remarks show the fallibility of memory, if they also illustrate Foch's peculiarly strong tendency to assume that the facts of a situation coincided with his conception of it.' For the prevention of a break-through was primarily due to the fact that the Germans were not aiming at one, and so did not exploit the gap actually made. Only three French divisions were brought from Artois and the first did not arrive on the battlefield until the 25th, so that the Germans had time to consolidate their hold.

In point of fact Foch's aim was not merely to 'close the gap', but to regain lost ground. For this purpose the reserves available to the Allies at that time were quite inadequate, yet so strong was Foch's personality, and so vacillating that of Sir John French, that the latter found himself leaving Cassel at midday on the 23rd committed to a course of

1. Liddell Hart, *Foch*, p. 177.

action exactly opposite to that which he had proposed. Sir John had gone over to explain to Foch that the situation of the British troops in the salient was now so precarious that steps in preparation for their withdrawal would, in the absence of an immediate and decisive French counter-offensive, have to be taken immediately. There is no record of the exchanges at this interview but the outcome was that French agreed to shoulder the main weight of the fighting, and counter-attacking, on the strength of some vague promises by Foch which a proper scrutiny of the Allied dispositions in the area would have shown to be impossible of fulfilment.

By the time that French got back to his headquarters all the British reserves[1] in the area had been brought forward and, in spite of the morning expenditure, strength on the new 'front' was greater than at any time previously. When the Commander-in-Chief's orders for a second counter-attack were received at 5th Corps H.Q., General Plumer replied that it should go in as soon as possible as 'the longer that the Germans are given to entrench, the more difficult it is going to be to dislodge them'. To this end it was decided to put in the whole of the 13th Brigade in the gap on the left flank, i.e. at the end of the shortest direct line of their forced march from billets to the front.

The men had had a meal earlier in the day and observers who watched them filing across the Brielen Bridge described their condition as 'cheery, but physically very tired'. The attack to which they were committed never had any prospect of success. It was directed, like that of the morning, against Mauser Ridge, but this time from west-south-west. Once again, though, the infantry had to advance in broad daylight over ground that was very open, broken only by a few widely separated hedges which sloped gently up, over a distance of about half a mile, towards the enemy positions.

1. These consisted of the 1st Cavalry Division (dismounted) and 13th Brigade of the 5th Division under Brigadier-General Wanless O'Gowan, consisting of 2nd K.O.S.B., 1st R. West Kent, 2nd Duke of Wellingtons, 2nd K.O.Y.L.I. and 9th London.

And once again this had to be done without any co-operation from the artillery.

The attack was originally timed for 3 p.m., but owing to the difficulty of getting the men into position by then—much less making even the most cursory reconnaissance—zero had been postponed an hour. Unfortunately the gunners, scattered as they were, coming under different commands and being fed from different telephone exchanges, were not informed of the change. As a result of this they opened fire for a short time at 2.45 and then for an hour the assembly of the attacking troops proceeded in almost complete silence.

The men's boots squelched in the mud, but they were too exhausted to speak, although onlookers were eager with accounts of the horrifying effects of the poison. There had been no time to put proper protective measures in hand, but instructions had been issued 'that the troops should hold wetted handkerchiefs or cloths over their mouths; if possible these should be dipped in a solution of bicarbonate of soda'.[1] As the men trudged up to the line they passed the crowded dressing stations. In many cases casualties from the gas were lying outside by the roadside where they had been moved to pass their last hours in the fresh air, and to make room for the wounded from the morning attack.

'The whole countryside is yellow—the battlefield is fearful; a curious sour, heavy, penetrating smell of dead bodies strikes one. . . . Bodies of cows and pigs lie, half decayed; splintered trees, the stumps of avenues; shell-crater after shell-crater on the roads and in the fields.'[2]

Finally at 4.25 p.m., an hour and a half late, the men prepared to leave this region of ill-omen. Each battalion had been allotted 500 yards of front and was organized in six lines. The attack was to take place on the same ground as that morning, and the result was no different. As they entered

1. O.H., 1915, I, 195.
2. R. G. Binding, *Aus den Kriege*, pp. 89–91.

the enemy sights the first two lines were cut down where they stood, but the darkening smoke and a dust cloud gave some protection to those that followed and they gradually worked their way forward up the slope, picking up the survivors of the morning attack in some places, and got as far as the enemy outpost line before being brought to a standstill. On the extreme right the Canadians, following almost exactly in the tracks of the morning assault, did manage to close with the enemy and regain some of the farms and outbuildings that had been fortified in advance of the main German line, but, by 7 p.m., as dusk fell, all movement came to an end. Over 3,000 men had fallen without ever having come to grips with the Germans. In the whole of the 13th Brigade there was now not one officer or one man surviving who had fought the previous autumn at Mons or Le Cateau.[1]

The evening was cloudy, without a moon, and by 10 p.m. the survivors had been pulled back to a line running along the trough of 'Colne Valley'. Organization of this ill-chosen position proceeded only with great difficulty owing to the confusion and mix-up of units and the loss of so many officers; the ground was completely water-logged so that it was impossible to dig for cover; and the distribution of rations was scanty and haphazard so that many of the men were now twenty-four hours without food. The verdict of the *Official History* on the whole operation could hardly be more damning—no ground was gained that could not have been secured, probably without any casualties, by a simple advance after dark, to which the openness of the country lent itself.

The co-operation from the French troops on the west side of the canal, which Foch had promised, never materialized. In fact, Sir John French's own doubts seem to have returned even before he got back from Cassel to his own H.Q., for on arrival there he had ordered up the whole of the 2nd Army reserve[2] to the 4th Corps area with the intention of mounting

1. O.H., 1915, II, 207.
2. These consisted of the 50th (Northumbrian) Division, the 10th and 11th Brigades of the 4th Division and the Lahore Division of the Indian Corps, from rest billets at Bailleul and Mérville.

a heavier counter-blow as soon as his dispositions would permit it. One cannot avoid the conviction that the doubts so evidently felt by the High Command about the results of these unco-ordinated and ill-prepared attacks make their ordering still less excusable.

It is true that Foch, after his meeting with French, went to see Putz to urge him to greater efforts. But Putz seems to have been as little impressed at this meeting as he had been by Foch's unrealistic instructions of the previous day. Putz was of stronger fibre than Sir John French and, owing responsibility only to the King of the Belgians, was less susceptible to political pressure. He had seen the front imperilled by the collapse and flight of two French divisions and felt little inclined to sacrifice the lives of the few remaining Belgian troops in hasty counter-attacks before French reinforcements arrived. Consequently Foch could get no specific assurances. None the less on the way back he called on French and told him Putz's hand was being 'massively strengthened'—although knowing that the sum total of these reinforcements was two infantry battalions and three batteries from the coast defences at Nieuport.

It was certainly unfortunate that the crisis had arisen at the exact point where the authority of the three commanders —Foch, Putz and French—overlapped. The result was that considerations of personal vanity and prestige led to much bloodshed that might have been avoided by a dispassionate consideration of the military principles involved.

6

The Dismissal of Smith-Dorrien

S-D has been very unwise and tactless in his dealings with General Putz. His messages are all *wordy*—his pessimistic attitude has the worst effect on his commanders.

Diary of Sir John French, 26th April 1915

SUBORDINATE only to Sir John French, the command of the British troops in this area was in the hands of Sir Horace Smith-Dorrien. Smith-Dorrien was a clever, sensitive and rational man. No other officer of equivalent seniority—with the possible exception of Sir Ian Hamilton—was his equal intellectually, and none could rival his ability in handling large numbers of men with economy and decision. The Expeditionary Force itself had been forged under his practical and progressive regime at Aldershot where he was C.-in-C. from 1907 until 1912, and had been saved from extinction by his decision to stand and fight at Le Cateau in August of 1914.[1]

But, throughout the fighting that followed the gas attack, Smith-Dorrien's ability was hamstrung by the repeated interference of French, who was himself subject to constant and contradictory pressure from Foch. Foch's attitude—alternately excitable and wooden—had a disastrous effect on the nervous and indecisive British Commander-in-Chief. At a time such as this a cool head and a firm grasp of basic military principles were essential. But Sir John's chief concern was twofold: not to lose 'face' with the French; and to avoid any withdrawal which would show significantly on

1. See p. 48, fn.

the small-scale maps published every morning in *The Times*. The acquisition of even the shallowest strips of ground had proved so costly a procedure that it was felt at the British Headquarters, as at the French, that the reverse must be true, namely that no price was too high to avoid 'yielding' territory. The concepts of flexibility and manœuvre in defence—that were to be so effectively put into practice by the Germans in the spring of 1917—had, on the Allied side at least, already been choked in the mud. Thus it comes as no surprise to find that, in spite of the day's losses, the confused and demoralized condition of many of the units, the extreme tactical vulnerability of their position and the complete absence of any effective protection against the new enemy weapon—in spite of all these things Sir John French was still thinking in terms of the offensive. By now it was plain that Putz was not going to lift a finger to remedy the situation, and as much was realized at G.H.Q.; but without moderating, it seemed, their views as to the proper manner to conduct the battle. In a message to Smith-Dorrien, timed 9.30 on the morning of the 24th, Robertson admitted that 'evidently not much reliance can be placed on the two French divisions on your left', but continued: 'The Chief thinks that vigorous action E. of the Canal will be the best means of checking the enemy's advance.'[1]

However, hours before this message was sent, the Germans had forestalled the British plans by a resumption of their own forward movement. They selected first the most vulnerable region, the sharp apex of the Canadian line at the eastern end of the breach, bombarding it mercilessly from 3 a.m. until 4.15, and then releasing gas. Sentries at first light saw the poison cloud coming on rapidly like a greenish fog bank, fifteen feet high, carried on the dawn breeze. The men had no protection but handkerchiefs, towels or cotton bandoliers wetted with water or any other liquid available in the trenches and little time to adapt even these pathetic devices, yet they clambered up and on to the parapet from where

1. O.H., 1915, II, Appendix 22.

they continued to fire into the advancing enemy until, as happened to over three-quarters of them, they were overcome by the gas. It is impossible to exaggerate the bravery of these Canadians, who, in their defiance of an almost certain and particularly ugly death, held on to these critical positions at the upper hinge of the Allied line. Gradually, as more and more of their number succumbed to the gas and fell into the pit of the trench, where the vapour lay strongest, their fire slackened. But the enemy had been so surprised at meeting any resistance at all after discharge that the mass of his attack had veered westwards and southwards into the open country below 'Kitchener's Wood', and he contented himself with an intermittent bombardment of the Canadian positions for the rest of the day.

Here was yet one more example of the gallantry of individual soldiers saving the commanders from the effects that should have followed on their own folly. For this whole, perilously balanced, triangle should have been evacuated during the lull of the 23rd, and the density of troops here redistributed along the crumbling face of the main breach. Had the German attack succeeded, as well it deserved to, the 28th Division line as far south as Polygon Wood would undoubtedly have collapsed with disastrous consequences. As it was, with the exception of a small inroad in the eastern end of the Canadian line, the enemy column was diverted across the sights of the defenders over the Keerselare crossroads and up the valley of the Steenbeek towards St. Julien, where it ran up against a scratch force of Highlanders, Buffs and some Canadian remnants,[1] supported by some mixed batteries of field artillery firing over open sights, and here it was temporarily checked.

By 5 a.m. wounded and gassed men were streaming back into the centre of the Salient, where the confusion was, if anything, worse than that of the evening of the 22nd. Although Brigade Headquarters had information of the new

1. Chiefly men from the 13th, 7th, 3rd, and 2nd Battalions and the survivors of the original night attack on 'Kitchener's Wood'.

attack at 4.30 a.m. and was already calling urgently for reinforcements, the 4th Corps was not told before a message sent at 7.20 but not received until 7.40. Thereafter there was no further news for General Plumer until 11.33, when a message came reporting the new Canadian line in the Gravenstafel–St. Julien area. Thus it was that the whole burden of conducting the fighting, which was now raging at close quarters along the entire front, fell on the battalion commanders themselves, none of whom could have any idea of the general position. It is plain from a reading of the *Official History* that the command structure, which cannot ever be said to have been properly established in the threatened area, had now disintegrated. The result was a series of local, limited, last-minute withdrawals punctuated by counter-attacks carried out with great bravery but little regard for casualties or relation to the general situation.

All this confusion and loss of life is directly attributable to French's insistence on 'vigorously checking', i.e. maintaining close and aggressive contact with, the enemy. Permission to step back and regroup on the 23rd, when the Germans had halted, would have enabled Smith-Dorrien to reassert his control over the 4th Corps without which centralized direction any serious counter-blow was doomed to failure. French's insistence on 'directing' the battle personally meant that with four generals (French, Plumer, Alderson and Smith-Dorrien) attempting to control some five brigades the command set-up was as confused at the staff level as it was in the field.

Students who have noted the C.-in-C.'s complaint, regarding Smith-Dorrien's 'wordy' messages, set out at the start of the chapter, may well consider the following example of French's own instructions to his army commander as being a good example of a text at the same time ambiguous, prolix, and obscure:

'The Chief does not want you to give up any ground if it can be helped, but if pressure from the north becomes such that the 28th Division *ought* to fall back from its line, then of course it *must* fall back, for such distance as circumstances necessitate. But we hope the necessity will not arise. The Germans must be a bit tired by now, and they are numerically inferior to us as far as we can judge. In fact there seems no doubt about it.' (The italics are those of Sir John French.)

In this way the whole of the 24th April passed in a welter of local counter-attacks and confused hand-to-hand fighting, with the enemy gradually pressing forward—largely owing to his co-ordinated direction and superior artillery support —into positions from which he successively levered off larger slices of the British line. The most serious inroad made was along the axis of the Poelcapelle–Wieltje–Ypres road towards St. Julien, which the Germans occupied during the afternoon. This had the effect of gravely accentuating the Canadian salient to the north-east, which had already been badly split by a gradual deepening and widening of the small breach made there by the dawn attack. The Canadians were thus forced to evacuate their positions, which they had striven so heroically to hold against the gas attack that morning, and to fall back under fire to a scattered and unconnected series of trenches across the neck of their old salient—the remains of an old French redoubt known as 'Locality C'.

Almost unbelievably, at the end of this day of carnage, Sir John French persisted in ordering the counter-attack originally planned for the morning. He had now collected, in the vicinity of Ypres or less than a night's march away, nearly 12,000 fresh troops, consisting of the 10th Brigade (Brigadier-General Hull), the 150th Brigade, and six other battalions scraped up from the reserves of other units in the 2nd Army. Smith-Dorrien had wished to use these troops to shore up the line between the canal and the Poelcapelle road, which was still very thin, being held only by the shattered remnants of the 13th Brigade and those units that had taken part in the unsuccessful counter-attacks of the preceding days, but both Plumer and Alderson supported French, and he was over-ruled.

Accordingly at 8 p.m. that evening Brigadier-General Hull, whose men had not yet crossed over the canal, received orders to mount an attack against the enemy positions in St. Julien and 'Kitchener's Wood' at 3.30 a.m. that same night, superseding his previous instructions which had been to relieve the survivors of O'Gowan's and Geddes' men behind Hill-top Ridge, and entailing a four-mile night march across unknown[1] country devastated by fire.

The men had a terrible struggle crossing the torn-up ground. Their progress was at right angles to the various tracks that pivoted out of Ypres like the spokes of a wheel, and the night was pitch dark with pouring rain; they had to make endless detours to avoid the burning groups of houses that stood round every cross-roads and drew fire all through the night; the soil was water-logged—deep mud clung to the boots and clothing; constantly, in the darkness, the men stumbled on mounds of rotting flesh—pigs, mules, cows, sometimes the bodies of French-African troops that had lain there for weeks. The whole area was intersected by old trench lines, many of them dating from the previous autumn and now fused with the broken-up drainage ditches to form long stagnant dykes up to eight feet deep. Everywhere were discarded pieces of equipment, empty shell-cases, blown-up G-wagons, coils of barbed wire and the small white wooden crosses that marked the 'combat' graves. Before setting off, Hull had sent a note to Alderson's H.Q. asking for postponement of zero from 3.30 until 5.30; but, just as in the attack of the 23rd, the news of the postponement was not conveyed to the artillery. So, between 2.45 and 3.15, the men, still stumbling patiently forward to get into position, heard the precious ammunition that was to have supported their assault being fired blindly off into the night.

1. When the brigade arrived at Brielen Bridge the guides who were to bring it across-country to the attack positions were not there, and after waiting for an hour the men set off under the directions of two officers from the 149th Brigade who had appeared there 'by mistake'.

Finally, as the sun rose behind the stumps of 'Kitchener's Wood', the 10th Brigade advanced. Once again, like the 13th Brigade two days before, they were being called upon to attempt the impossible. Without adequate artillery preparation or support, on ground unknown and unreconnoitred, they were being sent off to turn the enemy machine-gunners out of a position which had ready-made cover in ruined houses and thickets, and splendid artillery observation from higher ground behind. None the less, the brigade 'advanced in faultless order, worthy of the traditions of its home at Shorncliffe',[1] but 'they were mown down, like corn, the dead lying in rows where they had fallen'. Some of the men got to within a hundred yards of St. Julien before dying. The brigade was virtually annihilated, losing 73 officers and 2,346 other ranks in under two hours.

. . . . .

That evening Smith-Dorrien drove to Sir John French's headquarters at Hazebrouck in an effort to dissuade him from ordering any more attacks. He pointed out that the French contribution was quite negligible, and that the exhausted condition of the troops and the confusion of intermingled and depleted units made some sort of a pause for reorganization imperative.[2] Sir John, however, was steadfast in his contention that 'he did not wish any ground given up if it could possibly be avoided', and the situation should be 'cleared up', and the area 'quieted down', as soon as possible.[3] The French were making 'a big effort' on the following day, and it was his wish to support them to the fullest.

1. O.H., 1915, I.221

2. After the defeat of Hull's attack the command set-up was reorganized as follows. All troops to the east of the St. Julien–Wieltje road were placed under the command of General Bulfin (28th Division), those to the west up to the canal under General Alderson with directions '. . . to reorganize the command, putting battalions under their proper Brigadiers if possible, or at any rate under a definite General'. O.H., 1915, I, 248.

3. O.H., 1915, II, 16.

As Smith-Dorrien drove back to his H.Q. he passed near the huts at Ouderdom where the Lahore Division, the last major unit left in reserve, was arriving, as yet ignorant of the fact that it was to be committed to battle the following day. As the horses were watered and fed, little groups of Sikhs and Pathans—they had only arrived a fortnight ago from Hong Kong—stood about miserably in the mud.

That evening copies of Putz's orders arrived at 2nd Army H.Q. and these showed that the French attack was, in fact, to carry very much less weight than they had originally stated. Only one new division, less a brigade, was to be put in, together with those troops already in the line. Hardly had this news arrived than a fresh message came saying that zero was to be put forward from 5 p.m. to 2 p.m. Smith-Dorrien immediately telephoned Sir John and protested that the French attack was not only too light to have any effect but was being put in at a time which would once again involve the British forces taking part in an all-night march to their battle stations. However, no regard was paid to these points and he was instructed to proceed as arranged.

The task of the Lahore Division was exactly the same as that set the men of Geddes' force on the 23rd—namely a frontal attack against the German positions on Mauser Ridge. Although the attacking troops were more numerous this did not begin to balance the fact that the enemy had enjoyed an extra three days and nights in which to improve his defences. Indeed the artillery support was, if anything, lighter than on the 23rd, for the division's artillery had, although allotted its sites on the west bank of the canal, no time to lay lines to its observation officers or range the guns on to their targets. The Germans, on the other hand, had almost doubled their own artillery strength since that day as the majority of their field batteries had been moved forward into the captured territory and dug in to their new positions.[1]

Early and ominous evidence of this came soon after 10

1. 27th Reserve Corps (Reichsardiv.).

a.m. when the long column of marching British and Indian soldiers was spotted by observation planes and heavy-calibre shells from the enemy long-range guns began falling among them. The fire continued, gradually rising in intensity as the division deployed in the lea of Hill-top Ridge, causing casualties and retarding the organization of the assaulting lines. At 11.30 French's headquarters telephoned what the *Official History* describes as 'a sorely needed message of encouragement', stating that the enemy '. . . could not be very strong or very numerous, as he must have lost heavily and be exhausted'.[1]

In fact the British numbered over 15,000 for, in addition to the frontal attack on Mauser Ridge by the Lahore, there was to be a simultaneous assault on the right flank in the direction of 'Kitchener's Wood' by the 149th (Northumberland) Brigade, a completely fresh unit of North Country territorials;[2] thus the attackers enjoyed, like the Dervishes at Omdurman, a substantial numerical superiority—although their chance of success must be rated even lower than that of those naked, stone-age savages.

The battle—if the afternoon's massacre may be dignified by such a term—lasted three hours. The attackers were never able, in the words of the official account, 'to close with' the enemy. As soon as the leading lines of infantry breasted the skyline of Hill-top Ridge they came under a heavy and persistent fire from the German field batteries and, as they trudged their way patiently down the reverse slope and into 'Colne Valley', the first of the enemy machine-guns began to pick up the range with long feeler bursts. Shells from 5·9 in. howitzers fell among them with pitiless accuracy, '. . . knocking out whole platoons at a time; British and Indians were falling quite literally in heaps'.

The two British battalions had been placed on the flanks, with the Indians stretching along between them. As the men toiled up the slope towards the German wire, through a dark

1. O.H., 1915, I, 258.
2. In fact they were the first territorials to go into action as a brigade.

Donkey decorates Lion. General Alderson pins the D.C.M. on an unknown Canadian lance-corporal, for bravery at '2nd Ypres'

Pinned-down in the wire

'Rawly'

Scaling-ladders going up

'Walking wounded'

Aubers: looking back at the English parapet. Twenty-three bodies of the 2nd Cameronians within forty feet

Polished boots: *left to right*, Joffre, Poincaré, George V, Foch, Haig

'Tower Bridge' and German wire in front of Loos village through telephoto lens

Germans wiring the 'Second Position' (Lens road in the background)

twilight of brown smoke from the shell-bursts, they entered a veritable storm of machine-gun fire,

'very heavy and very well directed and probably owing a good deal to the yellow flags which had been issued to the leading sections with instructions that they were to be prominently displayed so as to show our positions to our own artillery'.

Several of the Indian battalions, broken by the casualties and the loss of all their British officers, disintegrated; some of the men remained crouching in shell-holes, others turned and made their way back to the old line in 'Colne Valley'. But the 1st Manchester, together with some Pathans and Sikhs on the left, and the Connaught Rangers on the right, managed to get up to the edge of the enemy entanglements. Then, at this critical moment, the enemy played their trump card: they had brought a number of gas cylinders up during the night on the forecast of a favourable wind and as the survivors of the British and Indian attack reached their wire the poison was released at the western end of the attack frontage and blew diagonally down on the attackers, with deadly effect. The assault, which had been faltering, now came to a dead stop.

On the extreme right, in the direction of 'Kitchener's Wood', the attack of the Northumberland Brigade fared even worse. Although the brigade was in close reserve at Wieltje, Brigadier-General Riddell, its commander, did not get his instructions until 1.30 p.m., that is, half an hour before zero. The brigade took another half-hour to get under way—and then was moving forward over practically flat, fireswept ground with little more idea of what was required of them than the direction of the attack pointed out on the map. By the time the men reached the British front line the main attack of the Lahore had already been stopped and Brigadier-General Hull, whose remnants of the 10th Brigade were the garrison there, and who had been instructed to co-operate in the attack 'with at least two battalions', refused. The grounds for his refusal were the complete absence of any support on the left and the rumours of gas which had already spread along the line. Doubtless he was mindful also of his brigade's experiences over that same stretch of ground the previous day. Undeterred, Brigadier-General Riddell led his men over the top to almost certain death. He himself was killed at 3.40 p.m. after seeing over half his men fall within a hundred yards of the British front trenches. The attack never got anywhere near the enemy, the leading troops being pinned down in No-Man's-Land all night and withdrawn the following morning. In this, its first action, the 149th lost 42 officers and 1,912 other ranks, or nearly three-quarters of its strength.

. . . . .

The sole contribution of the French to the day's attack had been a noisy and ineffective demonstration by some black troops on the front between Lizerne and Het Saas. These wretches, advancing with a 'stiffening' of whites who had instructions to shoot any man who turned in his tracks, broke when the gas was released and shot their officers. They fought their way back to, and through, their own lines and ran amok for hours in the rear area looting dumps and raping the nurses in the dressing stations. So disordered did the situation become that General Putz had to request the use of one of the British cavalry brigades from Vlamertinghe to assist in restoring order.

This note, coupled with the feebleness of the French preparations for the following day, as evidenced in the copy of Putz's orders which reached the 2nd Army that evening, decided Smith-Dorrien. Without consulting Sir John French, he issued orders to the effect that offensive operations were to cease forthwith, and that 'consolidation' was to be the object of all future dispositions. At the same time he despatched a strong note to Putz in which he protested that it was impossible to order any further attacks by the 2nd Army until there was evidence of substantial and effective co-operation by the French. He also sent a long letter to Robertson, French's Chief of Staff, setting out the position and going on to raise the question of a possible complete evacuation of the salient.[1]

By now the condition of the 5th Corps was very bad. The Germans had artillery observation of all the main roads leading east and north out of Ypres and movement across the canal was becoming increasingly difficult even at night. With each day that passed the number of casualties accumulated and wounded and gassed men lay everywhere, unable to move. There is a haunting picture of an advanced dressing station taken at this time. The detail is clear, emphasizing many questions that suggest themselves to an observer. The building itself has been hit by a shell. When? Before or after it was housing wounded? The operating theatre is inside; dimly, the M.O. and two orderlies can be seen looking out of the doorway. Standing about outside, some of them smoking, are the stretcher-bearers, glad of the excuse to hang around a little before going back to the firing line. Among them can be seen the padre, a stout captain—how many men already that morning have died under his blessing, moaning some final message to their dearest in England? In the foreground are many stretchers. A closer look reveals that these nearly all carry dead men or those, like the wretched being on the extreme right of the picture, whose wounds are so grievous that they have been given up for lost. One of these in the immediate foreground has been upset from his bier. Was he put down roughly or suddenly as the scream of an approaching 5·9 was heard? Or did he roll off in his death agony? With this exception there are no empty stretchers. As soon as the dead have had their personal effects and identity discs removed they are buried and the bearers return with the stretchers to the battlefield.

Of those in the 5th Corps who had survived unhurt all were completely exhausted; they were in many cases serving under strange officers and in amalgamation with other,

1. See Appendix 3.

equally depleted, units.[1] There was a brooding atmosphere of death about the whole region with its bleached, poisoned crops and sour smell of gas.

And if, tactically, the aggressive holding of the salient was costly and ill-judged, in a strategic context it was completely inexcusable. Experience at Neuve Chapelle should have shown that nothing but a carefully prepared offensive, requiring much time to organize, could possibly dislodge the Germans from the ground they had gained and had been methodically fortifying for several nights. Yet the French were organizing a great offensive in Artois, to the assistance of which Haig's 1st Army was heavily committed, and two days earlier Sir Ian Hamilton's expedition had landed at the Dardanelles. Any further dilution of the military effort was absurd and dangerous.

Yet the moment that Smith-Dorrien's letter arrived, Robertson telephoned back to say that 'The Chief does not regard the situation nearly as unfavourable as your letter represents',[2] and directing 'vigorous' co-operation with the French attack that was to go in that afternoon.

By now no one nearer than G.H.Q. to the front had any hope of success. Even Foch seems to have tired of the whole thing, for his promises of reinforcement had dwindled; his notes had taken on a sharper tone; he declared that the troops on the spot were sufficient—'*Pour poursuivre l'affaire et la résoudre*'. When zero hour came the French infantry never left their trenches and the remnants of the Lahore Division—the only 'reserve' left for the attack—were cut to ribbons in No-Man's-Land without even getting as far as they had the previous day.

That evening before the fighting had died down a message from G.H.Q. came into the 2nd Army switchboard 'in clear'

1. For example the 2nd Army order of battle, 28th April 1915, shows a 'brigade', in numbers hardly stronger than a full-strength battalion, with units from three separate brigades: 2nd D.C.L.I. (82nd) 260 strong; 1st York and Lancaster (83rd) 280 strong; 5th King's Own (83rd) 400 strong; 2nd Duke of Wellingtons (13th) 350 strong. Total 1,290, under the command of a lieutenant-colonel.

2. O.H., 1915, I, 397 (and see Appendix 3).

directing General Smith-Dorrien to hand over the command of all troops engaged round Ypres to General Plumer, and also to hand over his chief staff officer, Major-General Milne. Thus, although Smith-Dorrien was the senior of the army commanders, the effect of this order was to reduce his command to that of a single corps. For ten days he remained in this position while the very situation which he had foretold came about under Plumer's command—that is to say a gradual, clumsy, forced and costly withdrawal, or contraction, of the lines round Ypres. Even this was carried through half-heartedly so that, instead of pulling back to the town ramparts and the line of the canal, the British troops were left in a miserably cramped and shallow bulge, dug in on the reverse slopes of ridges that gave the Germans perfect observation of the whole area and 'a permanent target for artillery practice for the next three years.'[1]

Total British casualties in the salient for the period 22nd April until 31st May, the date when the new shape of the position was finally settled and the battle, known as 'the Second Battle of Ypres', officially came to a close, were 2,150 officers and 57,125 other ranks.

But before this, on the 6th May, Smith-Dorrien wrote to French suggesting that the evident lack of trust in him constituted a seriously weak link in the chain of command, and that for the good of the cause it would be better if he should serve elsewhere, and someone else command the 2nd Army in his place. That same evening he received written instructions from G.H.Q. directing him to hand over the army command to Plumer and return to the United Kingdom. No reason or explanation was offered.

1. An excellent impression of the conditions under which the infantry had to fight after the withdrawals is given in some extracts from the diary of an officer in the Leinsters over this period, reproduced in Appendix 4.

7

The Second Experiment: Aubers Ridge

. . . by means of careful preparation as regards details, it appears that a section of the enemy's front line defence can be captured with comparatively little loss.

From a G.H.Q. memo to officers of field rank and above, 18th April 1915

IN SPITE of the serious mishandling of the Neuve Chapelle offensive, and the fate of the counter-attacks at Ypres, the private papers and public utterances of the British commanders alike show a mounting confidence in the weeks that followed. The divisional commanders were agreed that a future operation '. . . should be Neuve Chapelle over again, but much more successful because we have learnt its lessons and shall know what to avoid this time'.[1]

But in fact, as the plan took shape at G.H.Q., it was plain that no lesson had been learnt, at least so far as concerned the direction of the battle. Tactically it was to be fought on exactly the same lines. Once again massed man-power was to be asked to overrun a thin firing line and, even if it was overrun, no measures had been thought out to counter the problem of the concealed machine-gun nests, 800 to 1,000 yards behind, that at Neuve Chapelle had caused such tremendous losses and demonstrated beyond all doubt their effective stopping-power. That the difficulties which that battle had shown might be expected when an attack by massed man-power enters a position defended by intelligently

1. O.H., 1915, II, 13.

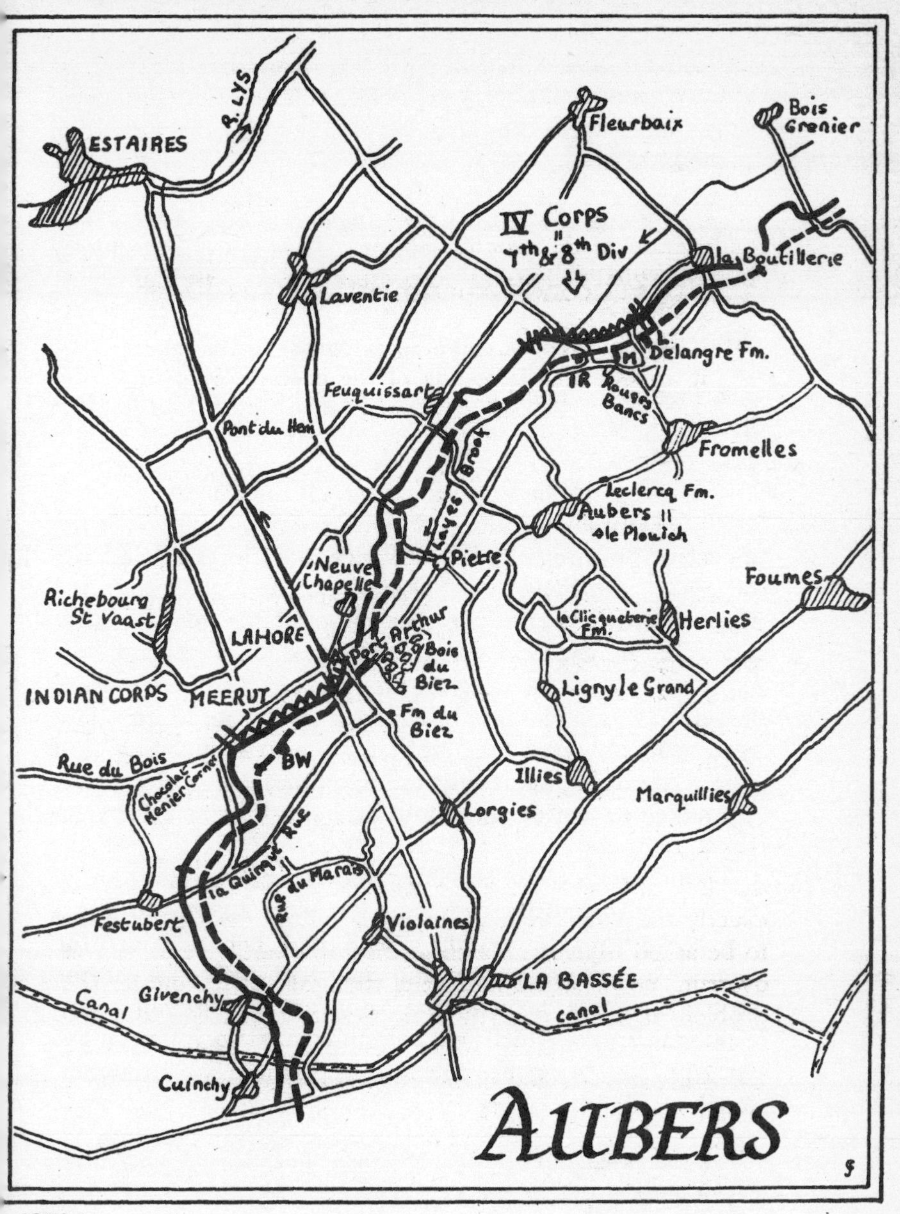

KEY

Attack frontage • Penetrations L London Regt.

M Munsters IR Irish Rifles BW BlackWatch (p.m.)

applied fire-power had not yet been considered. It is thus not surprising to find Falkenhayn summarizing the situation,[1] simultaneously to the day with the G.H.Q. memorandum quoted above:

> 'The English troops, in spite of undeniable bravery and endurance on the part of the men, have proved so clumsy in action that they offer no prospect of accomplishing anything decisive against the German Army in the immediate future.'

Before examining the conduct of this disastrous operation in detail, however, one must look at the planning that preceded it.

General Joffre had conceived the idea of a massive spring offensive, attacking along the front Arras–Lens, with the object of storming the Vimy Ridge and breaking out and beyond it into the plain of Douai. On the 24th March, less than a fortnight after the collapse of the Neuve Chapelle offensive, he wrote to French asking whether he would be interested in co-operating on the left flank.[2] Joffre was 'very hopeful . . . said he was bringing up even more troops and really thought he would break the line past mending, and that it might be, and ought to be, the beginning of the end. He talked of getting to Namur and the War being over in three months.'[3]

French conferred with Haig, who echoed the optimism of Joffre. Both the British commanders were anxious to dispel the amateurish impression which had been left by the Neuve Chapelle operation and they had evolved a scheme of attack, operating over the same terrain but more ambitious in scope, which was communicated to Joffre on the 1st April, as being 'feasible in approximately four weeks' time'.

1. *General Headquarters (1914–18) and its critical decisions*—English translation, 1919, p. 74.
2. O.H., II, Appendix 4.
3. Wilson, *Memoirs*, 125.

This time two sectors of the German breastwork were to be assaulted, on either side of the old Neuve Chapelle battlefield: a front of 2,400 yards from the Rue du Bois by the 1st and Indian Corps, and a front of 1,500 yards opposite Fromelles by the 4th Corps. After forcing two breaches in the German breastwork at those places, 6,000 yards apart, the two columns were to spread out and advance concentrically, joining up on the Aubers Ridge that lay about a mile and a half behind the German line.

It had originally been intended that the British attack should start a day later than that of the French, timed to hit the enemy as his reserves began to be drawn southwards, but owing to the differences in artillery preparation Sir John French insisted that his attack go in at the same time as that of D'Urbal, who was commanding the French 10th Army on his right flank.

This was reasonable for, once the element of surprise was lost, the British chances would be in serious jeopardy. Although the French, with 959 light and field guns and 293 heavies, were intending to attempt the passage of No-Man's-Land in the wake of a preliminary bombardment of five days' duration, General Haig, with rather less than half that number of guns, intended to rely on a bombardment of forty minutes, 'of which the final ten minutes was to be intense'.[1]

In so placing his faith once again in surprise, the 1st Army commander was influenced, doubtless, by the shortage of ammunition. But he was ignoring also the repeated warnings of the Intelligence Section and, indeed, the evidence that presented itself to the naked eye of any observer in the front line.

Neither of these sectors had in fact been touched by the earlier fighting, but ever since that time the Germans had been working ceaselessly at their improvement. And now, in May, their front line resembled more the huge sprawling earthworks of Sevastopol than the flimsy structure that had

1. O.H., 1915, II.

been shattered by the gunners eight weeks earlier. The front breastwork, built of sandbags and revetted with large-mesh wire, had been doubled or trebled in thickness to from fifteen to thirty feet across with frequent traverses and a parados.[1] In addition huge sandbag mounds, provided with shelters and a dug-out accommodation, had been built at varying distances—30 to 200 yards—behind the front for living in (*Wohngraben*). Then, as at Neuve Chapelle, there were a number of machine-gun nests (*Stützpunkt*) situated some distance behind the line to act as rallying centres in the event of a break.

Although the enemy front was lightly held the defence plan[2] was realistic, and thoroughly rehearsed. As ill fortune would have it it was based on the very principles most likely to wreck the British attack scheme—namely of blocking at once the flanks of a break-in by using the support companies on switch lines formed by the newly dug communication trenches. These, running at right angles to the front, connected with the *Stützpunkt* and had a fire-step on each side so that troops occupying them could shoot right or left. They also contained deep concrete shelters fitted with water pumps to which the garrison could hurry as soon as the bombardment began, leaving only a few sentries to watch at the front breastwork. The German troops themselves were sited in depth, each battalion holding its front breastwork with a garrison of two companies (each of 140 rifles), with one in support some 2,000 yards behind and another in reserve about two miles back.

Against these two German regiments, the 55th and 57th, Haig's 1st Army mustered three corps, the 1st, Indian and 4th. The attack was to be made in the first instance by, from south to north, the 1st Division (Major-General R. C. B. Haking), the Meerut Division (Lieutenant-General Sir C. A. Anderson), and the 8th Division (Major-General F. J. Davies),

1. The facts of the German defence are taken from Das Inf. Regt. 55, Schulz, 1928; Das Inf. Regt. Castendyk, 1936, and the analysis of these diaries in an article in *Army Quarterly*, Vol. 36, No. 2, by Captain G. C. Wynne.
2. 57th (Ger.) Regt. diary.

with three more divisions, the 2nd, the Lahore and the 7th, in close support.

The final plan for the offensive was explained by Haig to his corps and divisional commanders at two conferences held in Béthune on the 27th April and the 6th May. In essence the scheme was for the 1st and 8th Divisions to fan out after a quick break-in, holding the southern and northern flanks of the breach, while the Meerut and 7th Divisions converged on a big *Stützpunkt* situated at La Clicqueterie Farm, some one and a half miles behind the centre of the German front.

In theory, once these two units joined hands the whole body of enemy defending the unattacked portion of the line would be caught in a noose. However, in spite of the experiences at Neuve Chapelle, no definite objective for each day was given. Although many regimental officers would have preferred a limited one with 'systematic exploitation after the first assault', the course of the battle was left to a great extent in the hands of the local commanders. The attacking divisions simply had vague orders to press on as rapidly as possible from objective to objective to the line of the Haute Deule Canal, a distance of some six miles beyond Aubers Ridge.

During the nights of the first week in May there was great activity on the British front. A number of shallow disused trenches in No-Man's-Land were reclaimed as jumping-off places for the assaulting lines; blocks of assembly trenches were prepared a few hundred yards behind the main parapet where it was hoped that the second and third waves could shelter until their turn came; and both these new areas were connected up with the main positions by narrow communication trenches. But it proved impossible to dig these to a depth greater than two feet—a failing that was to have serious consequences on the day of the attack. At intervals of thirty feet steps were cut in the main breastwork to enable the infantry to pass over more rapidly and, in addition, the assaulting companies were provided with portable stepladders for this purpose as well as a number of light foot bridges that had to be carried forward for use in crossing the German trenches, the Layes Brook and other dykes and obstacles that might hold up the advance.

It was thus particularly unfortunate, in view of the congested state of the forward area, that the offensive had to be postponed at the last minute. Haig's final operation orders were issued on the 6th May at 10 p.m., naming zero as 5.40 on the morning of the 8th, which meant getting the troops into position the previous evening. But, on the morning of the 7th, there was a dense mist, limiting visibility to fifty yards, and worsened by intermittent rain and drizzle. These conditions upset the French bombardment schedule as the guns, which had been firing without a pause since the 3rd, could no longer range properly on to their objectives or observe the effect of their fire.

It was thus with considerable misgiving that the British officers heard to the north the continuous rumble of the French artillery die gradually away at the very moment their own attacking battalions were marching up to the line from their billets. As the day lengthened, uncertainty spread and rumour followed rumour. Finally, at five o'clock that evening, a letter arrived at Haig's headquarters from the commander of the French 10th Army,[1] explaining that the date of the attack would have to be put off until the 9th. All forward movement was immediately halted, but those men who had already got into position were kept there as it was felt that, in spite of the difficulties in feeding them and finding adequate space for the two extra nights for which they were to wait, any further movement would add to the confusion and might attract the attention of the enemy.

After midnight the French guns started again to the south, continuing their bombardment with mounting intensity all day on the 8th. But on the British front that day, and all

1. This was D'Urbal, for whose 'excessively good manners' Haig had earlier recorded his distaste (above, page 25).

through the night until dawn of the Sunday, the silence was complete.

. . . . .

The 9th May was a perfect morning of early summer. The larks could be heard singing as they circled to greet the rising sun. Then, exactly at 5 a.m., the British guns opened fire. The contrast with the tranquillity that had preceded the bombardment made it seem doubly impressive. Through their periscopes, observers in the British front line could see nothing but a high wall of smoke, dust and splinters rising from the German parapet, swirling and boiling as wire-cutting shrapnel from the eighteen-pounders exploded within it. But, measured by its actual results, the bombardment was seriously inadequate. Smoke and noise there were in profusion but the fall of shell was inaccurate, and their calibre in the majority of cases too light[1] to affect the massive German earthworks. The time allotted for the bombardment was quite inadequate for the few heavy guns that were available to work over the German position thoroughly. Moreover the effect was still further diminished by the fact that many of the shells were duds of American manufacture which had been filled with sawdust instead of explosive.[2] In the case of the 4·7 in. employed for counter-battery work these guns were now so worn out that as soon as the shell left the muzzle the copper driving bands stripped and the shell turned end over and fell anywhere, sometimes as little as 500 yards from our own support trenches.

Thus within twenty to twenty-five minutes of its commencement the German officers began to get their men out

1. The only guns capable of breaking up the enemy emplacements were the sixty-pounder and the 9·2 in. R.H.A. returns for 9th May show six 9·2 in. for the whole front (10th and 13th siege batts.) and twelve sixty-pounders (24th, 48th and Canadian heavy batteries). As for the wire-cutting eighteen-pounders —'the low trajectory of this gun at wire-cutting ranges, 1,500–2,000 yds., and the flatness of the ground did not make it an ideal weapon for the purpose; and its shells passing low over young troops and transport horses at night were trying to their nerves'.

2. 55th Infantry Regiment (German) diary.

of the *Wohngraben* and back into position behind the parapet. According to the British plan the leading companies of assaulting infantry were to go over the top and into No-Man's-Land at 5.30, that is at the moment when the bombardment entered its intense phase of the final ten minutes. But at about twenty past a strong breeze blew across the southern edge of the battlefield dispersing, momentarily, the clouds of smoke and dust; the German breastwork could be seen with alarming clarity to be almost intact while behind it showed the helmets and bayonets of men moving about. The brief duration, and vivid clarity, of this vision made it the more nightmarish, as any thought of holding up the attack was impossible to the junior officers and men who were its only witnesses.

Then, sure enough, some ten minutes later, as the first wave climbed out and over the British parapet, with the warm morning sun in their faces, the enemy opened a concentrated fire. In this region, the extreme right flank of the attack front, the assault was being led by the 1st Northants and the 2nd Royal Sussex of 2nd Brigade, and the Munster Fusiliers and 2nd Welch of 3rd Brigade, from General Haking's 1st Division. Side by side with them, on their left was the Dehra Dun Brigade, consisting of Ghurkas and Seaforth Highlanders from the Meerut Division commanded by Lieutenant-General Anderson, these two groups forming together the southern, or lower, arm of the pincer.

The Dehra Dun Brigade fared worst. The diary of the German 57th Regiment described how, as the bombardment lifted on to the rear areas and the smoke cleared, '. . . there could never before in war have been a more perfect target than this solid wall of khaki men, British and Indian side by side. There was only one possible order to give—"Fire until the barrels burst." '

As the German machine-guns scythed into advancing lines the confusion became intense; many of the men turned and made for the cover of their old parapet, but here they were met by the second and third waves who were attempting to climb out. In a short while the shallow jumping-off trenches were clogged with dead and wounded; the majority of units were in complete disorder having become intermingled with those following behind them, while many of their officers had been shot down[1] while standing on the breastwork exhorting the men to come out again and press the attack. In the official account it is said that 'the troops found it impossible to advance more than a few yards from the front parapet' but the German 57th Regiment diary does admit that a handful of Ghurkas got as far as the wire. They had discarded all their equipment, including their rifles, but 'running like cats' along the entanglement they found a gap and passing through attacked the defenders with knives. Alas, there was to be no recognition of this desperate gallantry for all were cut down and buried in a communal and anonymous grave by the Germans later that day.

On the right of the Meerut Division the attack suffered as heavy casualties and was but slightly more effective. Here, too, the artillery support was seriously deficient:

> 'For most of the batteries it was the first experience in wire-cutting and as only thirty minutes had been allowed the results were not unnaturally incomplete. Then, when the time came for the infantry advance the various artillery "lifts" were too quick—the first lift was made before the assaulters were within fifty yards of the Germans—with the consequence that the covering gunfire got clean ahead of the troops. As the telephone lines back from the front had been cut and no other means of communication had been arranged it was impossible to correct this.'[2]

Of the attacking battalions, the Sussex and Northamptonshire were practically annihilated in the passage of No-Man's-Land, suffering over a thousand casualties. But there was no thought of turning back to the cover of the trenches.

1. The 1st Seaforth lost all their officers in this way. (O.H., 1915, II, 23.)
2. O.H., 1915, II, 22.

In spite of losing nearly all their officers the men pressed right on up to the German wire and then searched along it —all the time under crippling fire—for gaps through which they could pass. But a cruel reward awaited them as it was found that a deep ditch that had been dug in front of the German breastwork for earth to fill the sandbags had itself been filled with tight coils of barbed wire. At this final obstacle the Sussex, now reduced to a handful, disintegrated; but a small party of the Northamptonshires, about twenty in number, managed to scramble across the ditch opposite a breach in the enemy breastwork and entered his trench, where they were at once engaged in desperate hand-to-hand fighting.

On the extreme right flank of the attack, also, a lodgement had been made. Here the Munsters, pressing forward with extraordinary bravery under the personal leadership of Lieutenant-Colonel Rickard—who was mortally wounded at the edge of the German wire—managed to penetrate with their right company and, after a fearful passage of the wire-filled ditch, the survivors of this detachment, also, entered the German breastwork. But by now the German fire was sweeping No-Man's-Land to such effect that the second and third waves had either been forced back into their own trenches or pinned down among the craters that lay between. Watchers in the British line could see, for a few moments, several of the Munsters standing on the enemy breastwork waving a green flag; then one of them was shot and the rest disappeared.

It soon became evident to General Haking that, with the exception of these two small entries by fragments of the Northamptonshires and Munsters, the attack had failed completely. More serious, the leading brigades were badly off-balance with their forward trenches choked with wounded, several of their companies pinned down in No-Man's-Land and their chain of command disrupted by heavy casualties among the officers. In spite of this, however, he ordered a renewed bombardment for forty-five minutes, to be followed immediately by a fresh attack at 7 a.m., and passed word along to headquarters of the Meerut Division requesting them to conform.

This decision was an unhappy one, and not only on account of the hopelessness of the task set the infantry. For large numbers of our own men, the majority of them wounded, were lying out in front of the German wire. Here they were ill protected from small-arms fire but were, at least, too close to his own line for the enemy to risk turning artillery fire on to them. But now, tumbled in craters, with little cover save what they could scrape in the dry earth of No-Man's-Land with their own bayonets, they were nearly all killed by this second bombardment, for the gunners had had specific instructions to concentrate on the wire as distinct from the breastwork. With horror those few of the attackers who had gained the relative shelter of the enemy breastwork watched their wounded comrades torn to pieces by the storm of shrapnel that played over the wire for the next forty minutes, itself emphasizing their own isolation.

When the time came for the second attack the stricken and disordered condition of the leading brigades made it impossible to achieve the same degree of co-ordination and numerical strength as had characterized the first. None the less a few officers managed to group together the remnants of various units and gallantly led these once more over the top when the bombardment stopped. All these individual groups, however, were cut down in No-Man's-Land within minutes of leaving their own trenches.

Thus it can be seen that Haking's use of the word 'dead-lock' to describe the situation in his report to the 1st Corps H.Q. at 7.20 a.m. was nothing if not optimistic. In the space of two hours he had dissipated to the point of annihilation two of the finest brigades in the British Army without achieving any material success whatever. None the less, he asked, in this report, if he could not commit the Guards Brigade which he had in reserve, stating that 'If the wire is cut by deliberate fire and more of the enemy's machine-guns are knocked out, the assault can be delivered again after midday.'

Before considering subsequent developments on this front, however, something must be said of the course of battle to the north where the left arm of the British attack was also engaged in bitter fighting.

8

Aubers Ridge: the Northern Attack

General Rawlinson: This is most unsatisfactory. Where are the Sherwood Foresters? Where are the East Lancashires on the right?
Brigadier-General Oxley: They are lying out in No-Man's-Land, sir, and most of them will never stand again.

Rifle Brigade Official History, p. 186

IN THE northern sector, the attack of General Rawlinson's 4th Corps was to be led by the 24th and 25th Brigades of 8th Division, operating on much the same plan as the 1st and the Indian Corps to the south. There were some minor differences in execution, however, and these, added to the fact that the 25th Brigade under Brigadier-General Lowry-Cole was made up of some of the most gallant and well-disciplined troops in the Expeditionary Force, gave the attack here an appearance, to begin with, of partial success.

A number of ingenious and unconventional means had been devised to break up the German emplacements. Two mines had been sunk and run under the enemy lines on the extreme left where the 1/13th London Regiment was to break through and hold the enemy communication trench as a switch line protecting the flank. The sinking of the shafts had been a perilous business, for the engineers had first to penetrate a thickness of fifteen feet of water-bearing loam before they reached the hard blue clay in which the tunnel was drilled. The tunnel itself was made only just wide enough for a man to crawl along, with 'lay-bys' at intervals of twenty feet, and each ounce of the tell-tale blue soil had to be carried away in sandbags. Proper ventilation of the tunnels was impossible and sometimes men with a load of explosive on their back would pass out unconscious en route, blocking the passage-way, and have to be dragged out by their feet over hundreds of yards. By the day of the attack, though, over 2,000 lb. of black gunpowder had been packed at the base of each mine.

Then, again to supplement the main artillery bombardment, a number of eighteen-pounders were brought right up into the front line the night before, using rubber-tyred car wheels to minimize the noise, and put in special emplacements from where they could fire into the German emplacements at point-blank range.

This supplementing of the barrage to some extent compensated for its short duration and lack of weight. On the extreme left, in particular, the attack went well forward in spite of heavy losses, and the London Regiment pushed right on over the line of the German third trench, taking the *Stützpunkt* at Delangre Farm in their stride, and then wheeling left as arranged, to protect the flank of the attack. In the meantime the other two battalions of the 25th Brigade, 2nd Rifle Brigade and 1st Royal Irish Rifles, had stormed the German breastwork in their sector and had pressed on to their first objective, the bend in the Fromelles road that lay some 200 yards behind the enemy front line. Unfortunately, however, portions of the German earthworks were so strong in this sector that the strength of the enemy fire had forced this attack southward into the areas where the close-support eighteen-pounders could be seen to have done their work properly. Thus a gap between the London Regiment and the Rifle Brigade, latent at the start of the attack, had widened to over 400 yards by the time they were established in the enemy position. Moreover, to the south, the attack of the 24th Brigade had been virtually snuffed out before even the men had managed to leave the assembly trenches, for in this area the German breastwork had been practically undamaged along its entire length; only a small party of Northamptonshires about thirty strong had managed to survive the passage of No-Man's-Land and fight their way into the German position.

Thus these three lodgements—that of the London Regiment on the extreme left, of the Rifle Brigade and Irish Rifles in the centre and the forlorn little group of Northamptonshires on the right—were all seriously isolated. And when Brigadier-General Lowry-Cole arrived at the front at 6.30 that morning it seemed that the situation had already deteriorated rapidly. The hard chatter of the enemy machine-guns could be heard all along the line—the more clearly as our own artillery had ceased firing. Through field-glasses, whole lengths of the enemy emplacements could be followed, lightly scarred but structurally intact except in those few places when the first rush of attacking infantry had broken in. No-Man's-Land was being continually swept by heavy fire and all forward movement had ceased.

But although they could get no further, the supporting battalions were still flowing up from the back areas into the confusion of the front trenches, that were jammed with stretchers bearing wounded, broken scaling-ladders, and large numbers of dead who had been dragged in off the parapet and lay in heaps awaiting identification and burial. The confusion was heightened by the presence everywhere of groups of leaderless men whose officers had been killed and who had been driven back to take refuge in a 'strange' section of trench.[1] The conditions for sending reinforcements forward were highly unfavourable and yet this had to be done unless the troops whose gallantry had carried them into the enemy position were not to be forsaken,

1. Disorder in the front trenches was always aggravated by this factor which accompanied the repulse of an attack. After their officers had been shot, the men would slowly work their way back to their own line, but, without proper direction in the smoke and noise and sameness of the cratered terrain, would often end up at a point some hundreds of yards above or below their original jumping-off point.

for with each minute that passed the pressure on them was increasing.

The most critical situation was that of the 1/13th London on the extreme left, for they, in obedience to their orders, had taken up a position that had originally been conceived as the flank of a clean break-through; this meant that they were facing due north, not east, and that they were especially vulnerable to enemy counter-attacks.

A succession of half-hearted attempts to reach them all failed until at ten minutes to seven two companies of the 2nd Lincolnshire broke into the German position just below the London flank. But once inside the complex of enemy trenches they became lost in the maze of sandbagged buttresses, and the Germans, more numerous and familiar with every twist of the line, reduced their numbers to a critical level in the twenty minutes or so of hand-to-hand fighting that followed the break-in. However, a junction of sorts was effected by one man, Acting-Corporal C. Sharpe, who fought his way up 250 yards of trench, using captured enemy grenades and a bayonet in his right hand.[1]

All efforts to spread the area of break-in to the south, however, were ineffective and the passage of No-Man's-Land was so difficult as to eliminate any prospect of reinforcement or relief of the troops that were still holding on in the enemy emplacement. Then, quite suddenly, watchers on the English parapet saw large numbers of men from the Irish Rifles streaming back across the front German breastwork and making their way back towards their own lines. They were under heavy fire from the enemy all the time and suffering severely—large numbers could be seen dropping as they ran. To add to the confusion, a body of German prisoners whom they were trying to bring back with them was believed to be an enemy counter-attack; rumours flashed round that the enemy was disguised in captured uniforms and a heavy cross-fire was opened from our own

1. He was awarded a V.C. for this. He was killed at Passchendaele in 1917.

trenches.[1] Brigadier-General Lowry-Cole was himself killed while standing on the parapet and attempting to rally the men and restore order.

By now the confusion within the British lines was so great that all movement was impossible. The extra communication trenches whose shallow depth had caused disquiet earlier were proving to be death-traps, and the normal system was so overloaded with stretcher parties and walking wounded as to be unusable. Furthermore, now that the attacks had been halted, the enemy artillery had lifted from No-Man's-Land and the assembly trenches and was systematically 'feeling' up and down the rear and support areas.

In spite of this, however, some individual runners from Haig's H.Q. were able to get through to General Rawlinson with 'urgent' instructions. These were that '. . . you must press the attack vigorously and without delay on Rouges Bancs'—i.e. frontally and in the same place. These orders were issued at 8.45,[2] and took about an hour and a half to be thoroughly disseminated. By the middle of the morning, however, it was plain to all the commanders on the spot that it was physically impossible to mount an attack with the shattered remnants of the assaulting battalions that remained in the front trenches, while the acute congestion in the immediate rear made the task of relieving them with fresh troops, and that of evacuating the large number of wounded that impeded free circulation, laborious and costly. It was plainly impossible to achieve a state of readiness before the afternoon. In consequence nothing was done to put these orders into effect.

As the morning wore on, however, Haig became impatient. The reports from the southern sector, from Haking and

1. The *Official History* offers no explanation of this débâcle, other than the following: '. . . exhaustive enquiry failed to discover any reason for the retirement beyond the fact that someone unknown had shouted the order "Retire at the double", which was passed rapidly along the line'. Of the Regimental histories written subsequently, that of the Irish Rifles makes no comment at all; that of the Rifle Brigade, whose flank was uncovered thereby, does no more than quote from the *Official History*.

2. O.H., 1915, II, 36, fn.

Anderson, had equally been disappointing, but with the difference that here both generals had shown a readiness to renew the attack. At 11.45 a.m., therefore, Haig issued further orders insisting that the attack should be pressed 'immediately'.

The arrival of these fresh and insistent orders caused consternation on 8th Division front. It was plainly impossible to get the two shattered brigades, the 24th and 25th, out of the way in time, and so Major-General Davies decided to improvise by ordering them to attack 'with what men you can muster' and the support of those troops that had managed to squeeze into position by that time—1.15 p.m. These were the 2nd Queen's, which had come across from 7th Division, and two Middlesex Regiments that had not been trained for the assault at all but were garrison troops who had already been in the line in that sector for twelve days and were intended to 'consolidate' after the attack.

Owing to the fact that the exact whereabouts of those few detachments of London, Rifle Brigade and Northants that were still holding on in the enemy position was not known, it was decided to concentrate artillery fire solely on that stretch of 500 yards south of the Fromelles road that had remained untouched. In other words it was proposed to assault the strongest part of the enemy position in isolation, as distinct from making a determined attempt to reinforce the small breaches already effected.

In fact, the attack was brought to naught before it even got started. The majority of the men were already shell-shocked and bewildered by their experiences that morning; they were serving under strange officers and with unfamiliar comrades; they knew only too well the strength of the enemy, in that sector above all. The preliminary bombardment was short, and more than matched by the fire which the German guns themselves put down on the assembly trenches. The majority of men never even climbed out into No-Man's-Land, although many companies were reduced by more than half as they huddled in the shallow, crowded forming-up places waiting for the whistle. By two o'clock the position had changed not at all, except that the 8th Division had suffered a further 2,000 casualties.

. . . . .

In the meantime Haig had arrived at Lestrem—headquarters of the Indian Corps and operational centre for the direction of the southern arm of the attack. The atmosphere at lunch was not an easy one, although the talk was mainly of horses and hunting, for General Willcocks had earlier been compelled to explain that the Dehra Dun Brigade had been so severely 'knocked about' that it was not capable of launching the third assault which Haig had earlier ordered for noon, and which he had been expecting to find under way on his arrival. Then, as the Indian orderlies were serving coffee, came more bad news: Brigadier-General Southey, commanding the Bareilly Brigade, reported that, owing to the congestion in the forward areas and the impossibility of using the communication trenches, his relief of the Dehra Dun Brigade had been attended by serious losses, was not yet complete, and precluded any possibility of getting the battalions into position for assault before 4 p.m. at the earliest. No sooner had this been digested than a further despatch arrived, this time from the north, from the 8th Division. In this the failure of the midday frontal attack, launched as a result of Haig's insistent orders, was reported, as was an estimate of the casualties suffered.

'. . . the Chief took it very hard. We had been getting reports all morning of how well the French were doing and he must have felt that they would be laughing at our efforts, as they did in December. He wrote something in pencil and handed it to the D.R. and left the Indian Corps mess without another word.'

While Haig was motoring from Lestrem to Aire, the position of the 2nd London, the Munsters and the Northants, still holding on inside the enemy lines, was becoming hourly more desperate. They had used up all their ammunition and were compelled to defend themselves with captured enemy equipment. Moreover they were suffering heavily from their own artillery which had been putting down intermittent fire ever since the failure of the midday attacks and which no longer appeared to be discriminating between 'captured' and 'enemy' stretches of the line. While the Germans would retreat to the *Wohngraben* the moment a bout of English shelling started, the British troops had no cover except the battered trenches where they lay, that had been reduced virtually to mud and rubble not only by the gunfire but by the ceaseless hand-to-hand fighting with hand-grenades that had raged over them all day. In an attempt to extricate themselves from this situation the Munsters, to the extreme south of the line, launched an independent attack still deeper into the enemy position. The sheer audacity of this move surprised the Germans and this small group, still bearing their green flag, carried the enemy's support position and broke out into the country beyond. But both the runners who had been detailed to crawl back across No-Man's-Land, to inform Brigade of this plan and ask for support, were shot, and the attackers were far too few in numbers for any possibility of success. The moment that the Munsters had passed through the German support line the garrison ran back down the trench, reoccupying it and, using the parados as a fire-step, poured a stream of fire into the backs of the attackers. A few hundred feet further on the Munsters were halted by a deep brook. Some of the men tried to swim across but barbed wire had been staked across the bottom and they were drowned. The survivors took up a position along the bank, but their position was now worse than ever and all except three (who were taken prisoner) were killed by the British bombardment that preceded the afternoon attack.

This, the assault fixed by Haig for 4 p.m., had a particularly inauspicious beginning. For when Haig got back to his H.Q. from Lestrem Brigadier-General R. H. K. Butler, the chief general staff officer who was always left in charge in his absence, handed him a despatch from General Rawlinson. The substance of this was that General Gough, commanding the 7th Division that had not yet been committed to action but faced the imminent prospect of this, had made a 'personal reconnaissance' of the ground. This had left him convinced not only of the 'uselessness' of putting in the 21st Brigade, but also of '. . . the certainty of any further attempt to attack by daylight being a failure'.[1]

It was an unpleasant position for Haig; for with the delays made by the Indian brigadiers fresh in his mind it must have seemed that his commanders were losing heart; only Haking, of the 1st Division, was still filled with 'the attacking spirit'. Moreover this refusal—it amounted to nothing less—of the 4th Corps to press the afternoon attack must obviously have serious consequences on the prospects of the 1st and Meerut attacks in the south, which would now be without any pretence of support from the northern arm of the pincer.[2]

And yet there was never at any time thought of cancelling the afternoon attack. The Germans, who thought that the British must have had enough punishment (*die Nase voll*) for the day, were amazed to see them coming across once again, in broad daylight, with the pipers of the 1st Black Watch playing, as the sun lowered in the western sky. And once again along the whole length of the attack frontage there broke out the harsh rasping stutter of the machine-guns; once again whole lines of men withered away, reduced to straggling mounds of twitching, agonized humanity. The Black Watch was the only battalion to get into position in time and so the only troops to go over the top at zero. They were alone in No-Man's-Land, little groups of kilted soldiers

1. O.H., 1915, II, 37.
2. The enemy had been steadily reinforced all morning, and reports to this effect were sent in both by artillery observers who could see the traffic and by the R.F.C. The 55th Infantry Regiment (German) diary gives the morning's losses as four hundred, but the three supporting companies had all moved into position by 2 p.m. so that the strength of the troops holding the front defences was actually greater in the afternoon than it had been in the morning.

trudging doggedly forward through the clearing smoke, drawn on by the wail of the pipes that could be heard a mile and a half away in Festubert. Miraculously, some fifty of their number reached the enemy position alive; and once inside found the enemy garrison in flight; standing on the parados they turned the German machine-guns round and on to the enemy, catching them in their own communication trenches as the retreating garrison ran into the reinforcements that were hurrying forward.

But like the Munsters in the north they were too few in numbers to exploit their success. The enemy, working round through other communication trenches, surrounded them and, throwing bombs and firing machine-guns from the traverses of the front breastwork which they had reoccupied after the Scots had passed through, destroyed nearly the whole party after an hour's bitter fighting.[1]

By the time the troops that should have supported the attack of the Black Watch were in position—some forty minutes late—those officers on the spot deemed it 'inadvisable', most fortunately, to put them in and they were kept waiting in the assembly trenches—a demoralizing experience for troops waiting for assault owing to the large numbers of severely wounded that crowded there, while others could be heard crying in No-Man's-Land.

Back in 1st Army H.Q. Haig had, by tea-time, drafted fresh orders to all units. These were that 'slow, deliberate fire was to be maintained on the enemy positions throughout the remainder of the day' and that an attack was 'to be pushed in with the bayonet at dusk'. But the report of his three liaison officers with the divisions concerned suggested that the commanders on the spot were so reluctant to press another assault as to seriously jeopardize its chances of getting started with any semblance of cohesion. In the light of these reports, therefore, Haig, at 6 p.m., cancelled the orders for the attack at dusk and travelled once again to Lestrem where

1. Corporal J. Ripley and Lance-Corporal D. Finlay were each awarded the V.C. for their conduct during this engagement.

he called a conference of the corps commanders and their senior staff officers. At this meeting Haig addressed his audience sharply. He considered the progress of the battle so far 'regrettable', he said, and insisted that 'results' should follow from the attack of the following day, for which he was allocating fresh brigades out of army reserve which was in his control.

But Haig's listeners can have been less than enthusiastic, for on their return to their units they set about collecting further evidence with which to dissuade him. When Brigadier-General Butler telephoned round to confirm their state of readiness at 11.30 p.m. that evening, Gough and Rawlinson suggested that, instead of being put in on the northern flank, the 7th Division should be brought round to the south to reinforce the 1st and Indian Corps—a move that must have meant cancellation of the attack as it could not possibly have been completed in time, and would have disrupted the whole front. This, and other reports of losses and the poor state of the units at the front, continued to pour in to Haig's H.Q. all night. The army commander thereupon summoned yet another conference—this time at 1st Corps headquarters—for 9 a.m. on the 10th. At this, however, the corps commanders were more vocal; earlier they had talked things over among themselves,[1] and now they harped on the shortage of ammunition, and on the worn-out condition of the 4·7 in. guns used for counter-battery work. In addition they had a fresh and spectacular excuse in an order from Kitchener that 20,000 rounds of eighteen-pounder and 2,000 rounds of 4·5 in. howitzer ammunition should be sent from France to the Dardanelles—instructions which had arrived the previous afternoon. They had news, too, of those few brave remnants that had been holding out on the other side of No-Man's-Land in the German position: all had finally been overwhelmed by the enemy, except for sixteen survivors of the Northamptonshires who had managed to return to the British lines under cover of darkness. After prolonged

1. O.H., 1915, II, 39–40.

discussion the corps commanders had their way, the attack was cancelled—'. . . it was generally agreed that the German defences were stronger than anticipated'[1]—and Gough's plan of bringing the 7th Division round to the south was adopted.

The losses of the one day's fighting that was the 'Battle' of Aubers Ridge were 458 officers and 11,161 men. It had been a disastrous fifteen hours of squandered heroism, unredeemed by the faintest glimmer of success.

But, in fact, more than heroism had been squandered, for the divisions broken on this day, like those at Ypres in the weeks before, were the last of the old regular British Army, that had the training and discipline of years behind them and whose musketry and 'fifteen rounds rapid' made German observers in 1914 think that there must be 'a machine-gun behind every tree'. Thereafter the gaps in individual units were filled first by brave but hardly trained volunteers, the 'New Armies' of the Somme; and, later, by the conscripts whose turn was to come at Passchendaele.

It is one of the great strategic ironies that 1915, the year of opportunity for the Western Allies while so many German divisions were tied down in Russia, was in fact used by them to blunt the very instrument needed for victory. It marked, too, a final ossification of tactical thinking; after Aubers Ridge surprise was abandoned. 'Weight of metal' was regarded as all-important; the 'war of attrition' was held to be the answer. In theory the artillery became the chief weapon of offence and the infantry the moppers-up. This in turn led to a complete neglect of infantry tactics. As will be seen, even at the time of the Somme offensive over a year later, the infantry were still being directed to advance in lines, 'dressing from the left' at a brisk walk, and forbidden, under pain of court martial, to take cover in any 'trench, hole, crater or dug-out'.

Prevailing military thought was summed up by Robertson: 'We are like a gambler who must always call his

1. O.H., 1915, II, 40.

opponent's bluff. Whatever chips he puts down, we must put down more.'[1]

But he forbore to state the obvious, that at cards all money thus staked comes back to the victor—in war lives are gone for ever.

1. This remark of Robertson, and the substance of the comment thereto, which I have paraphrased, are taken from Captain Liddell Hart's personal file on World War I.

9

Repercussions and Recriminations

Our attack has failed, and failed badly, and with heavy casualties. That is the bald and most unpleasant fact.

CHARTERIS, letters, 11th May 1915

BEHIND the lines, at General Headquarters of the French and British commands, and in the capitals, distance magnified the echoes of defeat.

The French offensive had not fulfilled the promise of its opening days and their losses, operating on a longer front, were likewise greater in proportion. None the less orthodox military thinking in both armies professed itself convinced that the 'attrition' method was the key to success. 'We lost some 10,000 men and never gained a yard,' wrote Wilson triumphantly in his diary. 'Now whose plans were right, Foch's or Haig's?'

But the French themselves had doubts. Poincaré tells how

'Colonel Penelon . . . leaves me no illusion whatever as to the Arras operations, which have utterly failed. The thing is over, the casualty list is very heavy, and we are not going to get through. In consolation I am told that the German losses are very much greater' [they were, in fact, less than half] . . . 'but this can be very little better than guesswork.'[1]

On the day that the offensive opened Castelnau had told him that he did not think that any very important result would be achieved, and, contrary to the opinion of Joffre, thought that they might have to look for a decision in some

1. Poincaré, *Au Service de la France*, Vol. II (translation Sir George Arthur, p. 137).

other theatre of war, Italy or the Danube. Ten days later Henry Wilson saw Castelnau and

'. . . found him very much opposed to attacks like this at Arras, which cost 100,000 men and did nothing except shatter four Corps. He is for big guns, lots of ammunition, deep entrenchments, wait for the English to appear, stop all attacks till some chance of real decision and so on.'[1]

It is strange to find such hard-headed sense among the senior generals of that period, stranger still to hear such sentiments from the lips of one who, before the war, had been regarded as the 'High Priest of the offensive' and whose influence had spread deep into the Army. But at that time Castelnau held no fighting command, and jealousy of Joffre and Foch may perhaps have helped to sharpen his logical perception. The junior French commanders on the other hand were only too keenly aware of the hopelessness of the offensive and, unlike their British colleagues, were more vocal in expression. Early in June Poincaré wrote that 'Everyone is complaining about Joffre, and especially about his entourage.' Clemenceau declared that 'if things go on as they are doing, there will be a revolt of the Generals against the High Command'.[2]

In response to a large volume of letters Poincaré visited the Arras front in July where he found ample evidence of the rift between 'brass and boots'. The Commander of the 9th Corps besought him: 'Pray, Monsieur le Président, do what you can to put a stop to these local offensives. The instrument of victory is being broken in our hands.' All the other commanders, with little variation, voiced the same opinion. But '. . . the Army Commander by no means agreed with them. His view was that if the troops are having such a bad time where they are, they should be pushed on up to the crest of the ridge.'[3]

1. Wilson, *Memoirs*, p. 233.
2. Liddell Hart, *Foch*, pp. 189–90.
3. *Ibid.*

In conclusion he noted that 'One gets the impression of profound disagreement between the man who sets the task and the subordinates who have to execute it.'

Unfortunately for French and Allied strategy, this bout of discontent was largely dissipated by a few well-selected reforms on the part of Joffre. The bribes that he distributed ranged from eight days' leave for all the N.C.O.s and men engaged in the Arras offensive to a reshuffle of the higher commands that gave him a new Chief of Staff and Castelnau command of the centre group of armies. At the same time he arranged for a regular sequence of conferences between himself and the other army commanders 'so that a unanimous front may be presented' thereby doing away, in name at least, with the duumvirate system of rule that had prevailed under himself and Foch. The power of promotion as a restorative of harmony was amply demonstrated at the first of these when Castelnau joined with the others in condemning the 'Eastern heresy', i.e. the theory that the Allies should stand on the defensive until British man-power and munitions had reached full tide, declaring that the English '. . . can pronounce at their ease, having no invaded provinces to liberate'.

All the same Castelnau still seems to have been uneasy at this subordination of the practical to the sentimental. After getting back from Chantilly he said: 'At present we have no plan, and we are like a cockchafer in a glass case; we keep on putting our heads out right and left haphazard.'[1]

. . . . .

The agitated condition of their Allies had done nothing to soothe the nerves and susceptibilities of the British commanders. Joffre was continually pressing Sir John French to keep up local offensives while his own attack corps were recovering their breath, and there were some tricky scenes:

1. Poincaré, *Au Service de la France*, Vol. II (translation Sir George Arthur), p. 147.

'Joffre got very excited and pointed out that we should be doing very little if we only relieved the 58th Division, and hinted at Government action, which luckily Sir John did not quite catch, and I got the chance of interpreting wrong; but as both were getting hot I got Sir John to go away, saying he would send an answer later.'[1]

Nor were feelings any more tranquil on the 'Home Front'. For some months past the Conservative Party had been chafing with eagerness to play a larger part in the direction of the war and the long succession of disappointments, the fact that no early end to the war was in sight, and, now, the fresh spate of rumours and accusations that spread with the return of wounded and leave-men after the Aubers battle—all these were giving rise to dangerous political undercurrents. The British commanders could not but be aware of these, and being convinced, naturally, of their own indispensability, took steps to secure their position.

Their excuses centred round those that had finally tipped the scales at Haig's morning conference of corps commanders on 10th May, namely—shortage of ammunition. This was, of course, a distortion of the facts for it was planning, tactics and leadership that had lost the battle on the first day. The real reason for halting the battle was that the men were exhausted and had already suffered crippling casualties.

'The answer to any such excuse from a general is that you ought not to enter into an offensive battle unless you have enough ammunition. The Commander who miscalculates is seriously to blame.

'I well remember a soldier and Member of Parliament at G.H.Q.—Captain Stanley Wilson, M.P.—describing to me in those critical weeks with great prescience exactly what form this policy of offensive-defensive would assume. "We have failed, we have lost many lives." This was the gist of the

1. Wilson, *Memoirs*, p. 227.

G.H.Q. case. "There may be a popular outcry. Very well, then let us concentrate it quickly on the home authorities." '[1]

In pursuance of this policy French sent Captain Guest and his military secretary, Brinsley Fitzgerald, to London, armed with a quantity of secret information on the supply situation. These two talked most indiscreetly to leading members of the Opposition as well as 'with any other M.P. who could be got to listen to them'. They also put before Bonar Law, the Conservative leader, and Lloyd George, who was the principal rival to Asquith in the Liberal Government, a carefully edited mass of correspondence on the subject that had passed between French's H.Q. and the War Office.

Their arguments were given added force by the publication in *The Times* of a series of articles exposing the shells 'scandal'—whose very content showed that they owed much to official sources in the field—over the name of Colonel Repington, that paper's military correspondent. Repington was much addicted to fashionable dinner parties—it was in this milieu that he had first made the acquaintance of Sir John—but it so happened that at that time he was in France, 'touring' the Western Front with Lord Brooke. Haig, who was apprehensive of enquiring journalists, refused to see them and did his best to obstruct all the reporters working on the 1st Army front at that time,[2] but the Commander-in-Chief took them in and poured out his grievances, supporting these with a mass of secret data and correspondence. (Evidently French did not scruple to suppress, or Repington to overlook, the despatch preceding the offensive in which the Commander-in-Chief had stated that supplies were 'adequate'—or his earlier decision to order ammunition in the proportion 75 per cent shrapnel/25 per cent high explosive, itself a most serious error of judgement when considered against the background of siege warfare which prevailed on

1. Lord Beaverbrook, *Politicians and the War*, Vol. I, p. 91.

2. On 21st May Haig wrote to C.G.S. 'recommending that no newspaper correspondent be allowed to come close to the front during active operations'. i.e. the duration of the war. (Haig, *Diaries*, p. 93.)

the Western Front.) Sir John harped on the ammunition shortage and made the most of the opportunity for airing all his private grievances concerning the Secretary of State.

He could not have selected a more timely occasion or a more receptive audience, for Repington was himself keenly aware of Kitchener's dislike and contempt for 'Social Soldiers' and had several times in the past attempted, at the instigation of Henry Wilson, to spread gossip in London to the effect that Kitchener was 'mad' and 'unfit to command a platoon' and so forth. Furthermore Northcliffe, Repington's employer, had just decided, as he confided in Lord Beaverbrook one afternoon in the Ritz Hotel, 'to go on attacking Lord Kitchener, day in, day out, until he had driven him from office'.[1]

Repington's first despatch appeared in *The Times* of Friday, 14th May, and gave rise to a major political sensation. Then, on the Saturday, the Government, already tottery, suffered a fresh and fatal blow. Lord Fisher, the First Sea Lord, resigned, having first written an anonymous letter in his own flamboyant hand to the leader of the Opposition, stating the fact of his resignation. The ostensible reason for his action was 'disagreements over the situation in the Dardanelles' but it was well known to be, in fact, the climax to a long series of disputes with the First Lord of the Admiralty, Winston Churchill. Churchill was, at this time (as in the 1930's), 'a person peculiarly odious to the Conservative party'.[2] Even had he so wished, Bonar Law would not have been able to restrain his rank and file if Churchill had stayed in office with a new Board of Admiralty beneath him, and all other considerations pointed to this being an ideal moment for the Tories to force their way into the Government. Accordingly Bonar Law addressed a letter to Asquith informing him that they would demand an early

1. Lord Beaverbrook, *Politicians and the War*, Vol. I, p. 95.
2. Charteris wrote: '. . . I have never had much belief in Churchill. He is so glib and his judgements seem always wrong. He has always the perfect explanation, like a child with the inevitable excuse that you cannot break down, but know to be untrue. (*G.H.Q.*, p. 95.)

debate on the conduct of the war unless he reconstructed his government. Asquith yielded, sacrificing both Churchill and Haldane, another member who was anathema to the Tories, and formed his new Coalition Government on 19th May.[1]

For the British commanders in France the significance of these Cabinet changes were twofold. In the first place, public and official attention had been effectively diverted from the Aubers débâcle. Scapegoats had been found and assurances had been given that reforms were in hand. Second, and more important in its long-term consequences, the 'Eastern' group in the Cabinet had been routed. Of the two men with real strategic vision at present directing the war policy of the Empire one, Churchill, had been dismissed; while the other, Kitchener, had had his authority restricted by Lloyd George's appointment as Minister of Munitions and the elevation to a separate department of what had hitherto been a committee subordinate to the War Office. While it was true that public feeling did not allow Northcliffe and Repington to pursue their victimization of Kitchener in the Press, his influence within the Cabinet itself had been sensibly diminished.

These political changes virtually sealed the fate of the Dardanelles Expedition—short of some spectacular and sudden military victory which was unlikely to occur without further quantities of men and supplies being diverted to that theatre. And it became plain that the Government must back the view that the war would be decided on the Western Front and, accordingly, to pay every attention to the opinions and demands of those in command there.

. . . . .

1. It is ironic that the most lucid account of 'The Shells scandal' and its political repercussions should have been written by Lord Beaverbrook (*Politicians and the War*, Vol. I, Chaps. VI, VII, and VIII), the man who was most responsible for seeing that an identical situation did not recur in World War II. In this he asserts that the fall of the Government was '. . . produced solely and entirely by the dissensions at the Admiralty'. This may be true as regards the *direct* cause, but, equally, there can be no doubt that the dissemination of 'facts' relating to the scandal played a significant part in conditioning public opinion to the need for a change and thus of dissuading Asquith from taking up the challenge and attempting, as well he might have, to fight out the issue on the floor of the House of Commons.

At the beginning of July two inter-Allied conferences were held, to soothe such differences as remained after the defeat in May and to co-ordinate plans for a new offensive. The first of these, a purely Anglo-French affair, was held at Calais on the 6th. At this were present MM. Millerand, Augagneur, Delcassé and Viviani, Mr. Asquith, Mr. Balfour, Lord Crewe (who was acting for Sir Edward Grey) and Lord Kitchener, as well as Joffre and French.

In spite of this no record of the conference can be found, and it is not mentioned in the French *Official History*. Poincaré in his *Memoirs* merely states that Millerand told him of an 'interview', adding 'a subsequent interview took place yesterday at Chantilly at which an agreement was reached'. Foch does not mention the conference in his *Memoirs*.

However, Asquith has given a diagram of the seating at the conference table in his *Memories and Reflections*, although he says little of what took place other than that '. . . the man who came best, not only linguistically but altogether, out of the whole thing, was Lord Kitchener'. It does seem, as is discussed in the following chapter, that at this meeting Joffre and Kitchener concluded a private agreement, so that they should know where they stood on the following day. At all events there was complete harmony at the military conference which took place the next day at Chantilly where 'Sir John French stated that he was fully in accord with General Joffre's views that the general strategic situation demanded the offensive, and pledged himself to the utmost of his means'.

It seems that, the day before, there had occurred, not for the last time in the war, one of those mutually agreed but almost unconscious deceptions of the politicians by the generals; for, in *World Crisis*, Churchill wrote:

'. . . the representatives of the Cabinet had argued against a further Anglo-French offensive in the West in 1915, and proposed offensive–defensive operations, and the French had agreed; General Joffre had agreed. The agreement was open and formal (but not recorded). No sooner had General Joffre left the conference than, notwithstanding these agreements, he had already resumed the development of his plans for a great attack.'

Whatever the means the result was the same and, with their governments acquiescent and the memory of the spring defeats receding, the spirits of the commanders in the field rose again.

As often as not this seems to have happened beyond the point of self-deception. Sir Edward Spears conveyed the atmosphere at headquarters:[1]

'General Rawlinson's bouts of optimism were apt to play ducks and drakes with the rigid economy of effort imposed on his army by the niggardly means at his disposal. One day he came to see Franchet D'Espérey and to my horror spoke as if he had unlimited artillery with which to support the French attack. He sailed in, a stick under one arm, waving the other, in a splendid humour due no doubt to the fact that he was going on a few days' leave which he was going to spend hunting the wild boar with the Duke of Westminster near Arcachon. His promises were quite unrelated to what was possible. I spent some difficult hours, tables of guns and munitions in hand, dispelling the hopes Rawly had so lightly conjured up.'

All the same the resources at the disposal of the British commanders increased significantly throughout the two months following the conference. In all, eighteen new divisions were sent out and grouped into an additional four corps. These were the first of the Kitchener 'New Armies', imperfectly trained it was true, but filled with enthusiasm . . . they were the very first of the volunteers of the winter of 1914.

One wonders what they would have felt had they been

1. Spears, *Prelude to Victory*, p. 88.

able to read an entry, in Kitchener's own hand, on a War Office memorandum at this time:

'The French have an almost unlimited supply of ammunition and fourteen divisions in reserve, so if they cannot get through we may take it as proved that the lines cannot be forced.'

10

Loos: the Plan

> Joffre and Sir John told me in November that they were going to push the Germans back over the frontier; they gave me the same assurances in December, March, and May. What have they done? The attacks are very costly and end in nothing.
>
> KITCHENER, memo to Robertson, 18th July 1915

WITHIN a month of the inter-Allied conference General Joffre, undeterred by his hard experiences of the spring, had approved further but no less grandiose plans. The ideal, he felt, was for two great convergent blows from the sectors Arras–Lens and Rheims.

These fronts were in fact too widely separated for direct tactical interaction of the offensives, being served by different railheads and systems, but Joffre ignored this and believed, as he expressed it in a memorandum to Sir John French, that:

> 'A successful break-through both in Champagne and in Artois was to be followed immediately by a general offensive of all the French and British armies on the Western Front which will compel the Germans to retreat beyond the Meuse and possibly end the war.'

French expressed his general agreement. He had to tread carefully with Joffre, he thought, for Henry Wilson was constantly at his elbow with fresh evidence of Kitchener's plans to replace him. There seems little doubt that French really did believe that Kitchener had proposed his replacement by Sir Ian Hamilton at the Dunkirk conference the previous November and that he had been saved only by the support of Joffre and Poincaré. Should he forfeit this for some reason, he would feel less secure.

But Haig, whose 1st Army was to conduct the operation, had other ideas. The last thing that he wanted was to be implicated in a sort of half-cock offensive with the French pulling all the strings. He made a personal reconnaissance of the area and found the ground 'for the most part bare and open . . . and so swept by rifle- and machine-gun fire from the German trenches and the numerous fortified villages immediately behind them that a rapid advance would be impossible'.

Glumly Sir John listened to his Army Commander. Although, less than a week before Haig's visit, when Foch and D'Urbal had taken him on a tour of the front, he had recorded in his diary, '. . . the ground which extends for some distance to the west of our trench line affords many advantages to an attacker', he now wrote that 'Our future plans are causing me a good deal of anxious thought', and that '. . . after a careful examination of the ground at Loos and Lens and a consideration of Haig's report I am doubtful of the success of an attack against these places, which I had arranged with the French to make'.[1]

Joffre, however, would not tolerate any objections. Abruptly he told the British commanders that 'Your attack will find particularly favourable ground between Loos and La Bassée.'

As Sir James Edmonds drily commented: 'He did not enter into any explanation of the reasons why he considered the ground favourable.'

A conference with Foch was arranged for the 27th July at Frévent, and at this Foch '. . . maintained it to be of vital importance that, regardless of the ground and strength of the enemy's defences, the British First Army should make its attack south of the canal in co-operation with the French'. This completely upset Sir John, who had also to cope with

1. E. G. French, *The Life of Field Marshal Sir John French.*

Henry Wilson, daily at his elbow with his assertions of the superior French military judgement.

Haig, however, remained adamant, while French shilly-shallied between the two, changing his mind from day to day; stroking his moustaches in perplexity as his Rolls tourer bumped and lurched over the roads between Doullens, Frévent and St. Omer. There were four such conferences during June and July and the date of the offensive was successively postponed from 'the first week in August' to 'the end of August', to 'the 8th of September', to 'the 15th September', and, finally, to 'the 23rd of September'.

At intervals Joffre himself was called in to deliver oracular pronouncement on the world-shattering importance of the operations planned. Both he and Foch gradually abandoned any pretence that the locale was favourable and were taking their stand on the ground that 'a fierce blow' was vital 'to the honour and prospects of the Allied cause'.

During this period Intelligence reports began to come in with disquieting regularity of the energy with which the Germans were strengthening their defences. It was becoming plain that they, at least, knew what was coming. They had perfect observation of the whole front of the assault: in the north from Fosse 8, a vast heap of slag and shale deposit from the mine workings that dominated the area and was virtually indestructible; in the south from the Loos pylons, a tall lattice-like structure of iron and steel that stood at the pit-heads outside Loos village. (The British artillery had been trying for months to destroy this erection, known to the troops as 'Tower Bridge' on account of the similarity in outline, but without success.)[1]

The general plan of the offensive definitely subordinated the surprise of the enemy to the methodical destruction of his defences. None the less it was disquieting that the French Ambassador in Rome should report that its delivery was a subject of general gossip.

1. After they had evacuated them, however, the Germans, with their heavier guns, brought the pylons down in two days.

After some further period the unfortunate Sir John French hit upon a compromise solution: might it not be possible to co-operate with Foch's attack 'by threat, and implication', i.e. without sending the infantry over the top at all? A 'storm of artillery fire'—he used this term freely, ignoring the fact that one of his and Haig's strongest objections to launching the attack was shortage of heavy guns—'a storm of artillery fire laid down for a period of days on the German positions would harass and destroy their forward elements, and lead them to believe that a heavy attack might follow at any moment'.[1]

The catch in this, from an ethical point of view, was that French could not quite decide when, or how, to explain to Foch, with whom he was co-operating, that he was not going to attack with infantry. Perhaps he hoped that he would not have to explain at all. His papers show scant attention to this problem. At all events it was not to arise, or not in that form, for Henry Wilson got wind of the scheme, and promptly communicated it to Pellé, the French Chief of Staff.[2] Joffre then sent a strong letter to Sir John saying that he expected him to attack with all his forces, and wished him to settle details with Foch. French, still eager in evasion, had a reply drafted that '. . . he would assist according to ammunition'. Wilson at once rushed to see Foch, who 'when I told him the story was quite open about the deplorable effect if we don't fight—Sir John had better walk warily'.[3]

Once again Joffre was appealed to. By now he was in a state of high indignation. The insubordination, for it seemed nothing less, of the British generals was threatening the whole structure of his offensive scheme. This was to be the attack that would finish the Boche in the west. Was it to be prejudiced by the timidity of these inexperienced British? It could not, it must not, be thwarted. Perhaps already he was feeling the first chill puffs of draught from the wind of popular indignation that was to blow him out of office a year

1. E. G. French, *The Life of Field Marshal Sir John French.*
2. Liddell Hart, *Foch*, p. 197.
3. Wilson, *Memoirs*, p. 245.

later. He appealed to Kitchener. Alarmingly he hinted that he, Joffre, was being held responsible for securing the 'full and proper' co-operation of the British, and that should he fail he would be displaced and the politicians would conclude a separate peace.

Confronted by this threat Kitchener travelled once again to France and visited French at his headquarters. 'K' was alarmed by Joffre's threat; he was still suffering from reaction to the disappointment at the Dardanelles. By now it was August, the 'black August' of 1915; daily there were reports of fresh disasters on the Russian front; he was tired and he felt his own reputation waning. And, added to all these factors, there is considerable circumstantial evidence of a secret agreement contracted some months before, between himself and Joffre.

At the Calais conference of 6th–7th July, Kitchener, who had accompanied Asquith, Balfour and Crewe in a strong delegation that was to settle Allied differences, had scored a great personal success in bringing the French round to the British point of view. The Cabinet were duly gratified at his achievements, but they were ignorant of the background. The facts of this are as follows:

In the early morning of 7th July, Kitchener took a stroll with Joffre, followed by a long private conversation in the saloon of the special train which had drawn the French ministers to Calais. Kitchener would never reveal any details of that conversation; but the main conference, which was resumed later that morning, ended in a complete agreement within a remarkably short time.

Joffre accepted, with apparent geniality, the decision to stand upon the defensive on the Western Front. But he took no notice of it whatever, and he continued to work secretly and uninterruptedly on his plans for a huge-scale autumn offensive in Champagne. He may have explained to Kitchener that the psychology of the French people made a defensive policy dangerous; and it is certain that Kitchener knew what Joffre was planning. He probably hoped that

Hamilton would break through the Dardanelles during August; at any rate the Cabinet reinforced Hamilton's force and he was given this last chance to succeed there.

But by the 18th August Kitchener was back in France with the knowledge that Hamilton had failed and that there was nothing left but to honour his compact with Joffre. On the following day he went to 1st Army H.Q. and, after addressing the corps commanders in the garden there, he asked Haig to see him privately.

'After washing his hands, Lord K. came into my writing-room upstairs, saying he had been anxious to have a few minutes' talk with me. The Russians, he said, had been severely handled, and it was doubtful how much longer their army could withstand the German blows. Up to the present, he had favoured a policy of active defence in France until such times as all our forces were ready to strike. The situation which had arisen in Russia had caused him to modify these views. He now felt that the Allies must act vigorously in order to take some of the pressure off Russia, if possible. He had heard, when with the French, that Sir John French did not mean to co-operate to the utmost of his power when the French attacked in September. He [Lord K.] had noticed that the French were anxiously watching the British on their left. And he had decided that *we must act with all our energy, and do our utmost to help the French, even though, by so doing, we suffered very heavy losses indeed.*'[1] (The italics are Haig's.)

1. Winston Churchill throws an interesting light on Kitchener's state of mind at this point in *World Crisis*, p. 463:

'. . . To avoid unnecessary circulation of secret documents, it had been arranged that the members of the War Committee wishing to read the daily War Office telegrams could do so each morning at the War Office in Lord Kitchener's ante-room. . . . On the morning of August 21st, I was thus engaged when the private secretary informed me that Lord Kitchener . . . wished to see me . . . He looked at me sideways with a very odd expression on his face. I saw he had some disclosure of importance to make, and waited. After appreciable hesitation he told me that he had agreed with the French to a great offensive in France. I said at once that there was no chance of success. He said that the scale would restore everything, including of course the Dardanelles. He had an air of suppressed excitement like a man who has taken a great decision of terrible uncertainty, and is about to put it into execution.'

It may be remarked that the reluctance of the British higher command to commit their infantry in hopeless or excessively costly operations stands in marked contrast to their attitude up to and following this date, and, in particular, to the profligacy with which lives were later squandered on the Somme and at Passchendaele.

The autumn offensives of 1915 came at a transitional stage in the development of the British armies in France. Hitherto the battles had been of an experimental nature; from a strategic aspect they had been—in spite of French's optimistic telegrams—holding actions. Under these conditions there was no room for indecision or 'faint-heartedness'. Commanders who wished to remain in office for the greater struggles that were to follow were not slow in taking to heart the lesson of Smith-Dorrien's dismissal; there was to be no squeamishness over 'losses'. But now, after a year of war, a change was coming over the scene. Soon Kitchener's 'New Armies' would be taking the field. The preponderance of British fighting strength over that of France, in quality if not yet in numbers, would be manifest. The prospect of honour and fame on a great scale were heavy in the air. To excitable minds there was even the possibility that the post of supreme commander, of Generalissimo, might ultimately be offered to an Englishman. How cruel to be robbed of such a prospect by involvement in a fiasco, at the bidding of an ally whose own position would shortly become in fact, if not in name, subordinate. To Haig, and to French, after Haig had explained the position to him, all these dangers were very apparent. His own personal position was not yet adequately consolidated. 'A new broom', 'a fresh start', 'cut away the dead wood' . . . these and many other spectres that haunt the leaders of a society, based on popular government and a popular Press, troubled them. If the attack should prove a failure, too obvious a failure . . .

But with Kitchener's intervention the burden was lifted from their shoulders. The responsibility was his, and that of Joffre. Moreover Haig was in touch with the King. As he looked about him he must have seen many who could be jettisoned if seas became too rough.

And now that the decision had been taken, Haig's own mind began to warm to the idea. He had built for himself an enormous wooden tower. From the balcony at its summit he could look out over the malevolent grey-black country with its slag-heaps, *crassiers*, and little clusters of miners' cottages, roofless from the perpetual shellfire but housing in their cellars the German machine-gun crews. Through his field-glasses he could discern the coils of barbed wire—in places thirty or forty feet across—that sprawled, like great poisonous centipedes, among the craters.

When an M.P., Mr. Shirley Benn, who was on a visit as a member of Lloyd George's ammunition committee, asked him how men could penetrate this wire Haig did not answer, but an aide told him that they would be cut by artillery fire. (Earlier another member of the commission, a Mr. McMaster, had asked Haig whether they still used 'the round cannon-ball'.)

The scheme that appealed to Haig at the present time was to attack behind a 'wave' of chlorine gas, projected from cylinders. He had attended, earlier, a convincing demonstration of the possibilities of this technique and it had, among others, the advantage that it allowed the widening of the front from a two- to a six-division assault.

There was a danger in this plan, however. To be sure of an effective 'wave' of gas, the engineers needed the wind in a certain quarter—west-south-west—and for it to be of at least moderate strength. Failing this the gas would simply hang about their own trenches, poisoning the troops crowded there waiting for the assault, or at best drift over towards the enemy lines in irregular gusts and patches, disrupting the uniformity and cohesion of the attack.

As some insurance against this Haig had an alternative, 'inner' plan for an attack on a two-division front if unfavourable weather should cause the larger scheme to founder at the last moment. But, again, this was already seriously compromised by the fact that he had spread his guns, in themselves hardly adequate to support an offensive on this scale, over the larger front, so that the concentrated attack would be starting with the fatal handicap of dispersed supporting fire.

Other factors, on a broader strategic level, were working against the success of the 1st Army's attack. Notwithstanding his earlier encouragement Joffre had privately become convinced, in the few weeks immediately preceding the date fixed for the opening of the offensive, that the ground in the Loos–Lens area was most unfavourable to the attacker, and had been shifting the main emphasis of his own armies to the southern stroke in Champagne. However, on 14th September he gave a final explanation of his plans at a conference at Chantilly, attended by the three army group commanders and by French. At this meeting Joffre declared that the time was 'particularly favourable for a general offensive', and expressed his 'confidence in a great and possibly complete victory'. The simultaneous attacks were 'a certain guarantee of success'.

1st Army Headquarters, too, were by now infected with optimism. Those who had doubts wisely held their peace, for 'disloyalty' or 'lack of offensive spirit' did not go unnoticed or unpunished.

It was generally felt that the gas would work wonders—a view which ignored the fact that although gas will poison men, regardless of nationality, only high explosives will destroy wire.

11

Loos: the Assault

Foch: *Est-ce-que les hommes sont en bon état?*
Haig: They never were in better heart, and are longing for a fight.

At tea, 12th September 1915 (Haig, *Diaries*, p. 103)

AT A quarter past five on the morning of 25th September, after an uneasy night spent in constant consultation with Captain Gold, the R.F.C. meteorological officer, Haig gave the orders to 'Carry on'.

Ponderously, for he had suffered a mild attack of asthma the previous evening, he climbed up the stairs of his wooden tower, his staff at his heels. As the sound of the bombardment, which had been unimpressive even at its height, abated, they peered across No-Man's-Land at the flickering bracelet of fire caused by the exploding shells as they crept slowly from the leading to the secondary German positions. So still did the air seem that, as the minutes passed, all became infected by the fear that the gas would simply hang about the British trenches. After a quarter of an hour Haig made one of his staff telephone to 1st Corps to enquire whether it was possible to stop the arrangements for the attack. The answer came that 'General Gough did not consider it practicable to get word in time to the front trenches'.

Nor were they the only ones in doubt about the wisdom of releasing the gas; in Horne's 2nd Division the officer in charge of the gas on the 2nd Brigade front declined to assume the responsibility of turning on the cylinders. On this being reported to Horne he ordered that 'the programme must be carried out *whatever* the conditions'. The reluctance of the corps and divisional commanders to sanction any last-minute alteration in the plan is all the harder to understand when one discovers the very complex and thorough arrangements that had been made to ensure a last-minute cancellation if this should prove necessary. Between the higher formations three routes were arranged, by telephone, telegraph and despatch-riders. To pass the order on to the gas units, officers, attended by runners, were stationed at special points. Each of these officers had ready twenty typewritten slips, 'Attack postponed, taps *not* to be turned on until further notice'.[1]

However, in spite of definitely unfavourable conditions on several parts of the attack front, these precautionary measures were nowhere implemented and, at nine minutes to six, the taps were opened.

. . . . .

There was certainly no shortage of gas. Until zero-hour at 6.30 over 150 tons were discharged from 5,243 cylinders concealed in sandbagged bays in the fire-trench. As the greenish-yellow chlorine came hissing out it slowly built up into a cloud from thirty to fifty feet high that billowed sluggishly forwards into No-Man's-Land. Overhead the German distress Verys curved red and white in the lightening sky and their forward machine-guns in the sap-heads began to chatter, firing short warning bursts at alternating elevations.

As the sun rose the wind did not increase. There can have been few among the infantry, packed like animals along the narrow slippery communication trenches, sweating in their improvised talc and flannel 'respirators', who did not feel a sense of foreboding as they waited for the subaltern's whistle. Far from being 'in a panic' the Germans had already begun to open bursts of deterrent fire and the bullets

1. O.H., 1915, II, 171.

were slapping into the sandbag parados just above the heads of the waiting assault troops. Soon mortar-fire was added to this and, further back, the enemy field artillery began to come to life.

In front of Loos and further north in the region of the Hohenzollern Redoubt the gas cloud carried fairly well over the German trenches and was to exert a marked influence on the advance of the 47th, 15th and 9th Divisions, only falling short of complete success because it moved too slowly and there was not enough of it. But at the southern end of the front the vapour, after thirty-five minutes' flow, was still short of the enemy parapet. And in the centre, on both sides of the Vermelles–Hulluch road, it drifted in the right direction at first; but towards the end of the discharge began to float back and into the British trenches, giving rise, in the words of the *Official History*, to 'great inconvenience and some loss'. In other places, particularly on the 2nd Division front, the discharge had to be discontinued at once and no gas reached the German trenches.

There was some surprise effect, but it quickly wore off. The official narrative of the German 6th Army reads: 'In general the physical effect on the men was trifling.'

The drizzle of rain had cleared, leaving a thin ground mist, when, at 6.30 a.m., the infantry clambered out of the trenches, and in the fog of gas and smoke, which made it difficult to pick up landmarks, began the advance across No-Man's-Land. They were in fighting dress—without greatcoat and pack, but cumbered with bombs, picks and shovels, and extra rations. All ranks wore the original pattern smoke helmet—a flannel bag—over their heads, but with the front rolled up, and had a second helmet in their haversacks. With the front down they could hardly see through the talc-covered eye-holes, and with the front up the rain caused the chemicals in the flannel to soak out and irritate the eyes.

Although casualties were heavy at every point these varied from mere decimation to whole battalions being virtually obliterated, as did their achievements vary from the substantial and heroic to the utterly negative. If the course of the battle, and the causes of the ghastly massacre of the following day, are to be properly understood, it is best to follow briefly the fortunes of each of 1st Army's six divisions, starting at the southern end of the attack front opposite Loos village itself.

The 47th Division, at the extreme southern end, broke cleanly through the German first line—the men of 1/18th London Regiment dribbled a football in front of them as they crossed No-Man's-Land—at a cost of some 1,200 casualties, or roughly 15 per cent, in the first hour. Unfortunately, however, the right and centre of the division halted at the German rear support trench, which they began to organize for defence, instead of pressing forward towards Cité St. Pierre. In the meantime the London Irish, on the left of the division, had entered the southern outskirts of Loos village and the 1/20th London, passing through the Irish, carried all before them, taking in quick succession the cemetery, the 'garden city', arriving still full of fight, though now sadly depleted, at the heavily defended 'Chalk Pit copse' by 8.30 a.m. An hour later they had fought their way into the pit itself but their numbers were too few to evict the Germans dug in round the copse and they suffered severely under enfilade fire from this quarter while their comrades looked on from the old German support trench some 800 yards away.

On the left of the 47th was the 15th Division, whose assault brigades were made up entirely of Highland regiments. At zero-hour their assault was seriously impeded by the obstinacy of the gas cloud which simply hung about the congested trenches. Many of the men lingered in the hope that it would disperse or drift away towards the German lines and there was much difficult to-and-fro traffic in the crowded fire-trench as platoons made their way to places clear of cylinder bays. The situation was saved, however, by the extraordinary heroism of Piper D. Laidlaw of 7th

K.O.S.B., who rallied the men by marching up and down the parapet playing 'Scotland the Brave' on the pipes, regardless of gas and enemy fire. He continued to play even after being wounded and was awarded the V.C. Once the assault got going the Highlanders pressed it with great vigour and complete disregard for losses. It took them less than an hour to penetrate both German trench lines in front of Loos village and by 8 a.m. they were enthusiastically digging the garrison out of the cellars at the point of the bayonet.

Unfortunately the enemy's fire, and the prospect of his rout in the village itself, had drawn all the Scottish regiments into the maze of trenches and connected cellars there, to the detriment of the broad plan of advance. Thus the front of the divisional attack narrowed from 1,500 to less than 600 yards and the 7th K.O.S.B. on the extreme left, who had achieved the deepest penetration of all, reaching the line of the Lens road by 9.15, were left in isolation, suffering intermittent shellfire from their own artillery. In Loos itself the Highland regiments were by now thoroughly intermingled. A very large proportion of their officers had been killed and many of the subalterns remaining did not like to assume the responsibility of giving orders, believing that their superiors were still alive but perhaps lost in the confusion. The men themselves—'a magnificent Border rabble'[1]—believed that it was all over bar the shouting and, by half past eight, were streaming out of the eastern end of the village in great spirits and starting the ascent of Hill 70 in a somewhat leisurely manner. They had, in the words of a battalion diarist, 'the appearance of a bank holiday crowd'. Furthermore, as they advanced up the bare slopes of Hill 70, the German garrison in the redoubt there, which was at that time no more than a few maintenance men and engineers, took to their heels. The sight of their enemy running away was too much for the Scots and with a renewed cheer they pressed forward and over the crest.

1. R.H. *Black Watch*, 1914–19, Vol. II.

But once they were on the downward slope the troops of the 15th Division were in full view of the Germans waiting behind the wire of their very strong second line, which had been built outside the range of the eighteen-pounders. And after they had travelled some half distance down the bare slope, fire was opened by the enemy. The Scots were completely pinned down. With only a few inadequate entrenching tools they could make little impression on the hard chalky soil. Some of them tried repeatedly to rush the wire. Others attempted to make their way back over the crest. But of the nine hundred or so who had advanced from the redoubt scarcely one survived.

During the day the Germans were rapidly reinforced and by the afternoon they were counter-attacking in sufficient strength to recapture the redoubt. The remnants of the 46th Brigade, now reduced to a handful, were rallied by Second Lieutenant Johnstone,[1] R.E., and made five separate attempts to retake it but were beaten off in each case.

The casualties of the division in this one day's fighting were nearly 5,400—or about 60 per cent—and some battalions, in particular the 9th Black Watch, 8th Seaforth, 7th Cameron, 7th K.O.S.B. and 1st Highland Light Infantry, were virtually annihilated. All the same the 15th and 47th Divisions had, though checked now and sadly depleted, made substantial gains in the first few hours. But further north, for the 1st and 7th Divisions, the situation was very different.

The attack plan of the 1st Division was, from the outset, of doubtful promise. On their right, or southern, flank No-Man's-Land was very wide. The opposing trench lines ran along the slopes of the Grenay Ridge, unobserved by each other and separated by the blind hump of the Col de Grenay on which stood 'Lone Tree', the enormous flowering cherry that had blossomed that May.[2] In consequence artillery observation both for cutting the wire and demolishing

1. Lieutenant Johnstone was awarded a V.C. for his part on that day. He was killed at Delville Wood, 1916.

2. After the blossoms had fallen a young lieutenant in the Seaforths had led

advanced saps had been very difficult, as also patrolling by night to investigate results. It had, accordingly, been decided to leave this sector out altogether and to concentrate the attack along the axis of the Vermelles–Hulluch road in the north, with the 1st Brigade to lead the attack and the 3rd Brigade in close reserve behind it.

At a later stage, however, as the plan worked its way up to Corps and Army level, amendments were made. In particular it was ordered that the Division's 2nd Brigade should after all make an attack to the south of Lone Tree. These instructions had the effect of weakening divisional concentration along the main axis of advance and, as the two brigades were from the outset directed to advance on diverging lines, threatened to aggravate this condition later. (To 'fill' this gap a composite force—known as 'Green's Force' from Lieutenant-Colonel E. W. Green, its commander—was created by taking away a battalion from each of the 1st and 2nd Brigade, and putting them back into reserve.) In particular, the orders meant that the 2nd Brigade was doomed, in effect, to be 'expended', for it had been given a task that was almost impossible.

And, as it turned out, the attack was a complete failure. The men were late in jumping off, as they suffered particularly from their own gas in this sector, and were badly enfiladed by machine-guns in two sap-heads that the Germans had run forward into No-Man's-Land. By the time the leading battalions[1] reached the wire they had suffered over 400 casualties including their commanding officer, Lieutenant-Colonel Sanderson. The wire itself was some ten yards across, firmly staked low in the ground, and

a night patrol there and, climbing to the upper branches, had attempted to fasten a Union Jack to the trunk. Unfortunately, although successful in this, he had been caught in a flare on the way down and machine-gunned. For several days his body had hung there. Two attempts to recover it on subsequent nights failed and finally divisional artillery were directed on to the tree in an attempt to bury him. As the days wore on all the branches had been blown off but the guns never scored a direct hit and the stump remained, standing some fifteen feet high. It flowered again in 1920.

1. 2nd K.R.R.C. 1st Loyal North Lancs.

virtually intact. Thus, within an hour after zero, the remnants of the 2nd Brigade found themselves pinned down in hopeless disconnected positions among the craters and depressions immediately in front of the main German position, their numbers being steadily reduced by short-range artillery fire and mortaring. From time to time little groups would attempt to clear a way through the obstacle with wire-cutters, but all were shot. By 7.45 the smoke and mist had cleared, for the British artillery fire had long since passed on to more distant objectives, and the prospects of an assault became still more hopeless.

At this point the Brigade may be said to have disintegrated. 'A few officers,' the *Official History* records, 'nevertheless rallied their men for another effort; but the attackers had lost heart and, individually and by groups, began to struggle or crawl back to their original trenches.'

The setback here, and the influence that it exercised on the mind of the British commanders, was to have very serious consequences on the operations further north.

. . . . .

Initially the attack of the 1st Brigade, opposite Hulluch, had prospered, although once again at tremendous cost.

There were two small copses in No-Man's-Land in this sector, known as Bois Carré and La Haie. For many months the artillery of both sides had passed over them reducing them to little clusters of shattered scrub and 1st Army Intelligence had classified them as 'unoccupied'. But in the weeks immediately preceding the offensive the Germans had run saps out and into the undergrowth here. The preparatory bombardment began along a line further in advance, and the machine-gun nests there escaped untouched. Their crews might have been incapacitated by the gas if things had gone right but, as it was, the cloud progressed so slowly that the three lines of British infantry were all deployed, fifty paces between each, and advancing in full view of the enemy, before the Germans smelt the first whiffs of vapour. Thus fire from these positions caused very heavy casualties before the attackers had even got to grips with the enemy.

On the right of the Brigade front the Gloucesters were the assaulting force. With extraordinary courage they forced their way into three successive German positions, but by the time they had penetrated the German support line and reached the maze of communication trenches that lay behind they were, in the words of the *Official History*, 'destroyed as a battalion'. The fighting was desperately exhausting and there was the utmost difficulty in keeping a proper cohesion to the attack. By the time that they arrived at the German wire the attacking infantry had, in almost every case, lost a proportion of their officers and N.C.O.s so that many sections were without proper instruction. The barrier itself was seldom penetrable along its entire length and platoons and companies would become badly intermingled and the confusion more serious as they searched for and passed through such gaps as existed. Once in the German lines it became even harder for those in command to keep a full control of their men. The enemy system was very intricate in this sector and the trenches, eight feet deep with a raised fire-step on their western side, turned back and forth every eighteen feet or so in a series of orderly, buttressed, right angles. At intervals steps would lead down under the parados to the dug-outs where little groups of unharmed Germans lurked ready to emerge with grenades and machine-guns after the first attacking wave had passed over, or where, more often, lay numbers of shell-shocked and badly wounded defenders suffocating from the gas that lingered there, inert and deadly.

In this evil-smelling maze the British infantry became still further dispersed, and it was only with great difficulty that they could be rallied and induced to clamber out over the parados and attack, once again over open fire-swept ground, the German support trenches that lay some eighty yards in the rear. None the less, within half an hour of first entering the trench, the subalterns of the Gloucesters managed to mount a second attack on the German positions beyond. This, too, was successful, though at a sad price. As the men advanced across the broken, cratered earth, whole platoons would be reduced to mere handfuls of individuals as the German machine-gunners scythed into them again and again. But at the last moment their extraordinary courage broke the spirit of the defenders who turned and fled down the communication trenches to Hulluch, leaving their guns silent and smoking on the parapet and the British to cover the last fifty feet unmolested. By now, though, the 10th Gloucestershire existed in name only; less than sixty, of all ranks, survived the first two hours of the assault.

On the left of the Gloucesters the attack fared as well, and was less extravagant in life. It had been rehearsed for weeks before by the Berkshires—the leading battalion—against replicas of the German trench system constructed behind the lines from aerial photos. The result of this thorough training was a clean break-through to 'Gun Trench' by 8 a.m.—a penetration of threequarters of a mile. This was the cleanest break on the whole front of the offensive, and that most urgently requiring exploitation. Here on 'Gun Trench', a shallow, wandering communication trench that connected a series of mortar pits, but was of little defensive significance, the Berkshires halted while the reserve battalion, the Cameron Highlanders, came up. Through the smoke the poplar trees along the Lens–La Bassée road could be seen. Immediately in front of them the firing had abated. It must have seemed that they were nearly through.

On their arrival the Camerons continued the advance at a good pace and by nine o'clock their forward elements had actually entered the village of Hulluch by progressing up Alley 4, a long 'arterial' communication trench that ran from the outskirts of the village across the northern part of the Loos Valley to the gun positions that had lain immediately behind. The German troops in the forward positions, never numerous, had been killed or wounded—there were many gassed and lightly wounded infantry lying on the floor of Alley 4 as the Camerons picked their way along—and the remainder had withdrawn, in considerable confusion, through the village and well behind the 'Second Position' for which, in spite of its natural strength, there were not enough men at that time.

Thus it was that the Camerons found themselves passing through 'gates' in the German wire, which the defenders in their haste had omitted to close, and heard their footsteps ringing in the deserted streets undisturbed by anything more lethal than an occasional shell from their own artillery, that was meant to be 'bombarding' the village. At the far, or eastern, end two enemy machine-guns and some infantry discouraged too close a follow-up without reinforcement but even they, in the words of the Company report sent back by the Camerons to 1st Brigade at 9.10 a.m.,[1] 'appear to be retiring'.

. . . . .

Here then, three hours after the start of the assault, was the critical point on the Loos front. For this small mixed force of the 1st Division—the Berkshires and the Camerons, and the remnants of the Gloucesters—were astride the German 'Second Position' at its most vulnerable point—that is, where it was closest to the original front line—with the choice of rolling it up to the north or the south, depending upon the course of the battle in those areas.

It was now essential to make sure that this spearhead could be adequately, and promptly, reinforced. Immediately available were the reserve battalions of the 1st and 2nd Brigade (Black Watch and 2nd Royal Sussex) and, less than an hour away, Colonel Green's force and the 3rd Brigade in its entirety—a total of some 6,500 men of whom none had yet seen action that day. Such numbers were more than adequate to force a clean break through the confused and

1. O.H., 1915, II, 213.

battered German elements that held on to the eastern end of Hulluch village, and open a way, at last, for the cavalry that stood patiently among the copse and scrub on the far side of the Grenay Ridge.

But speed was essential. With every minute that passed the German defenders had time to recover their composure; the reinforcements that had been directed there as early as the previous evening began to arrive; the guns were manhandled into their new emplacements; the infantry were assembled, given their orders, shown their field of fire.

For the attackers, this of all times was not one to worry about the flanks. Although, in fact, the extraordinary heroism of the attacking infantry had more or less secured these at every point except on the 2nd Brigade front on Lone Tree Ridge—which, anyway, it had originally been planned to omit from the attack plan on account of its strength. But it was, most unfortunately, this very position with which the 1st Division Commander, Major-General A. E. Holland, was preoccupied. It was incomprehensible to him that British infantry should be stopped dead, as the 2nd Brigade had been. He knew from Intelligence reports that the force opposing them must be a small one. It was now, furthermore, cut off from any prospect of help from either Loos (by the 15th Division's advance) or from Hulluch (by his own 1st Brigade). Another attack would surely bring about its surrender and the 1st Division's front would be 'clean'. With this in mind the two supporting battalions, instead of being directed to reinforce the 1st Brigade, were ordered up with instructions to clear the German position on Lone Tree Ridge and 'press on'.

As might have been foretold by anyone inspecting the situation on the spot, this second attack, made without any pretence of artillery support, in broad daylight, with no protection from smoke or gas, was cut to pieces.[1] More serious was the fact that, on the assumption that it would be

1. The gallantry with which it was pressed can be judged from the fact that three V.C.s were won on this short front during that same afternoon.

successful, Green's force had been ordered forward to fulfil their originally conceived—but now quite meaningless—role.

This had two results. In the first place the men opposite Hulluch village were deprived of the prospect of immediate tactical reinforcement from the 2nd Brigade reserve and, secondly, they saw the only substantial force (other than divisional reserve) that could have rendered them decisive assistance diverted to an objective that was militarily quite futile.

As Green's force set off it found that it, too, was under very heavy fire after breasting Lone Tree Ridge, owing to the failure of the latest attack by the 2nd Brigade reserve. Unable to use the communication trenches, which were filled with gassed and wounded men going in the opposite direction, both battalions were compelled to approach over open country under intense fire from a quarter which they had been told had already been successfully attacked. Soon they began to come up with the remnants of the 2nd Brigade and they, also, found themselves pinned down in the long grass in front of the German wire, unable to go forward or retire.

In the meantime precious hours were slipping past for the 1st Brigade, as the troops that it needed so badly for reinforcement were thrown away in frontal attacks directed against an enemy position that had already been outflanked. Lieutenant-Colonel Graeme, in command of the 1st Cameron Highlanders, could hear intense firing well to his rear, as the successive 2nd Brigade attacks went in. He realized the possibilities of an advance down the axis of the Lens road to outflank the enemy instead of repeatedly assaulting this front, and sent a succession of messages urging this course as well as that of occupying and consolidating Hulluch. But without reinforcement it would obviously be dangerous to further disperse his small mixed force and so they held on anxiously to the straggling cobweb formed by captured German gun-pits and such improvised trenches that the exhausted infantry had scratched in the hard chalk since their arrival.

Opposite them, the first of the German reserves were already beginning to arrive and move into the defensive belt that stretched away to the north and south, empty and undamaged. On their side, too, there was considerable confusion; the 26th Regiment, ordered up from Pont à Vendin, reported at midday: 'There appear to be no German troops ahead on a front of about three miles, and the forward batteries have all been over-run. How far the enemy has advanced is not known. The battalion will advance till it meets the enemy and be prepared for any eventuality.'[1] At intervals the men of the 1st Brigade would, from their advanced position, be presented with splendid targets as the enemy infantry, all unknowing, would march up in close order. At one moment the Camerons opened fire on a detachment estimated at over 300 that was proceeding down the road between Hulluch and Benifontaine, with great effect. But with each bout of firing it was plain that the enemy was becoming more numerous and the British ammunition less plentiful. The men were short of water, also, and gradually, as the hours slipped past without relief or contact with the units on their flanks, an ominous sense of isolation began to envelop them.

There were now only two units left on the divisional front that were available as reinforcement. These were the Brigade's own reserve battalion, the 1st Black Watch, and the divisional reserve of three battalions in the 3rd Brigade. If these forces had been sent up immediately their combined strength ought still to have been sufficient to 'prop open' the breach in the German 'Second Position' at least for the twelve hours or so that must elapse before the Army Reserve, the 11th Corps, could arrive on the scene.

At this point in the battle the situation for Brigadier-General Holland was, the *Official History* tactfully records, 'full of difficulty'. However, in spite of the open breaches to

1. 11th Division, War Diary.

the north and south of the German position, he decided against any outflanking movement and ordered yet another frontal attack, committing the whole of the 3rd Brigade and with it the last hope for any substantial help for the men in Hulluch. These orders arrived two hours late, owing to the loss of three runners in No-Man's-Land, and Colonel Green read 'with horror' the clear instructions to put in one battalion on either side of Lone Tree, to attack once again over this stretch of ground where the corpses were so thick and the groaning and calling for stretcher-bearers so insistent that the sound was 'like the cattle market at Devizes'.

Although he realized, as did everyone on the spot, the ease with which the German position could be outflanked, the orders were quite definite and, in view of the delays that had already taken place, there was no time to refer the question back to General Holland. So, at one o'clock, the two leading battalions (London Scottish and 1/9th King's) were sent over the top, the majority of them to certain death, for

'. . . the approach of another attack did not have the expected effect on the resisting power of the Germans. Before the advancing lines had reached the wire, still intact, they were greeted with a hail of bullets at close range. Every attempt to get into the enemy trenches was in vain, the men being shot as they endeavoured to cut a way through the wire.'[1]

The situation was aggravated by the fact that, owing to the sparsity of troops to the north, and their forward situation, the Germans on the 2nd Brigade front were all the time gradually working their way forward and northward along the trench line that the Berkshires and Gloucesters had passed over earlier in the morning. By midday they had even got a machine-gun back into position in the Bois Carré, which had been left unguarded in spite of the profusion of British troops in that area, and this had the effect of drawing off the 1st

1. O.H., 1915, II, 217.

Black Watch—the last available unit that could have reinforced the 1st Brigade at Hulluch—who were instructed to dig in and seal off the old German line at the junction of the 1st and 2nd Brigade fronts.

Thus, by the early afternoon, there was a state of deadlock along the whole of the 1st Division front. The offensive had lost all momentum; the men were exhausted, without reserves, had suffered fearful casualties and the strength of the Germans opposite them was increasing hourly.

At four o'clock the German force that had held up the 2nd Brigade, and all but destroyed it and Green's force, surrendered. But this was for reasons quite unconnected with the succession of frontal attacks to which they had been subjected during the morning. For the Germans had been taken at last—and quite accidentally—in the rear by a small group of the 2nd Welch from the 3rd Brigade that had been driven northwards by the fierceness of the fire and, finding themselves more or less lost in the wide shelving expanse of the Loos Valley, had worked their way down and back towards the sound of battle along the Germans' own communication trenches, taking the defenders by surprise.

By now, though, it was too late, for almost at the same moment a German counter-attack was driving the remnants of the 1st Brigade out of Hulluch and, although it was not pressed in sufficient strength to compel a withdrawal further than the line of the Lens road, its success did mean that the enemy 'Second Position' had now been restored in its entirety.

As this last short engagement died down the noise of battle abated. Leaden clouds, heavy with the rain that was to fall that night, darkened the Loos Valley as the remnants of the 1st Division trudged their way forward, unmolested now, their backs to a No-Man's-Land of hideous memory. Only the howitzers, eight miles in the rear, kept up their rumbling fire as the first big raindrops broke on the packs and helmets of exhausted infantry.

12

Loos: the Second Day

The machine-gun is a much over-rated weapon and two per battalion is more than sufficient.

HAIG, in a minute to the War Council, 14th April 1915

BEHIND the assaulting troops was the newly formed 11th Corps consisting of the 21st and 24th Divisions,[1] the first of Kitchener's volunteer 'New Armies', who had only arrived from England a fortnight previously. It was to this force, under General Haking (promoted to corps commander following the 'aggressive spirit' he had shown at Aubers Ridge), that Haig, now desperately short of troops, turned his eye.

They had spent the three nights prior to the battle moving up towards the line from their concentration area west of St. Omer and were in no condition to face immediate action. Moreover they were the only units in reserve behind the 1st Army front and Sir John French did not wish to see them used in the offensive. Nevertheless, as a result of Haig's urgent requests, the 21st and 24th were finally placed at his disposal.

Whether French intended them simply to consolidate the ground gained, serving as replacements for the enormous losses that had been suffered on the 25th, or whether they were to be used as an instrument with which to renew the offensive, is not clear. The real intentions of the two commanders have been obscured by the bitter controversy and

1. In the 11th Corps there was also the reorganized Guards Division but it was situated further back and separate from the 21st and 24th Divisions.

recrimination that followed on the fate of the two divisions. But there is no doubt that Haig and French diverged at this point. Haig saw his offensive already stalled. Unclear orders, fumbling at brigade and divisional level, and the enormous casualties that had followed thereon had seriously impaired the balance and condition of the attacking forces. The most that Haig could hope for was that the Germans were in similar plight. Perhaps he felt that another 'punch', thrown quickly, might still give him a chance to let loose his horses. The thought seems to have occurred to him that with these troops their very 'freshness' might be an advantage; with the enthusiasm of ignorance they would tear their way through the German line. Of them he wrote that 'having been so short a time in France they have not yet acquired the sedentary habits of trench warfare'. . . . At its crudest, they didn't know what they were up against.

But French seems to have been getting uneasy about the prospects of the attack even some days before its actual launching, and this may explain his half-hearted effort to keep the 11th Corps under his wing. On the 24th September he wrote: 'In view of the great length of line along which the Army is operating I feel it to be necessary that I should keep a strong reserve under my own hand.' Twice he resisted Haig's insistence on being granted absolute control of the 11th Corps, until finally he relented after visiting advanced H.Q. at Lillers at midday on the 25th.

Then, at the shortest notice, the 21st and 24th Divisions were ordered up from their billets—which were a considerable distance from the firing line—against a tide of congestion. As they made their way forward ugly rumours spread from mouth to mouth. Past them, in the opposite direction, the ambulances creaked and jolted in endless procession; among them, following the same routes and accorded priority,[1] were the convoys of fodder for the

1. Owing to the short notice at which the 21st and 24th Divisions had been ordered forward, no marking tapes or other arrangements for directing them had been provided along the route.

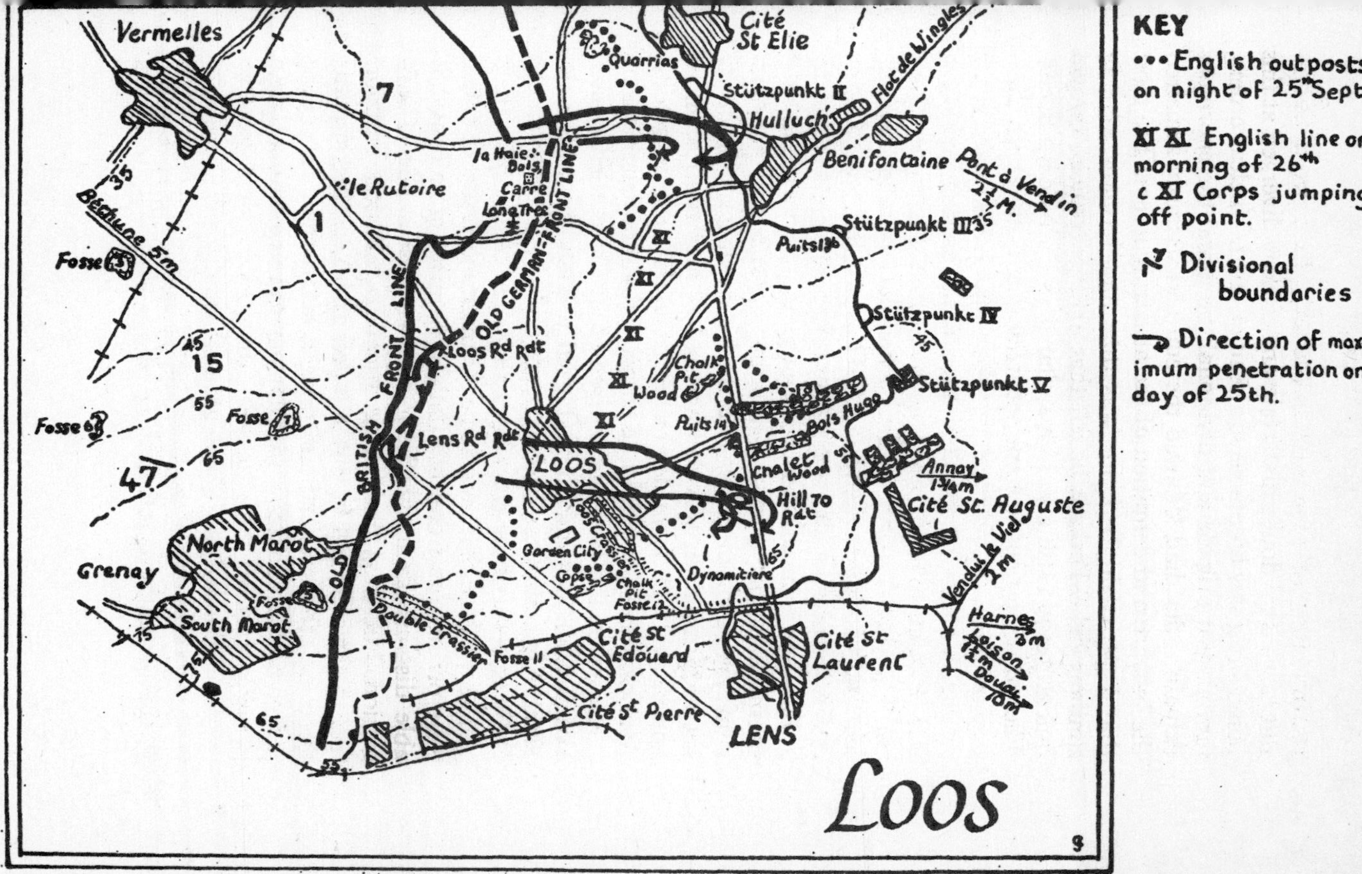
Vermelles
Cité St Elie
Quarries
Stützpunkt II
Hulluch
Flot de Wingles
Benifontaine
Pont à Vendin 2½ M.
Stützpunkt III
Puits 13 bis
Stützpunkt IV
Stützpunkt V
la Haie Bois
Carre
Lone Tree
le Rutoire
Béthune 5m
Fosse 3
Fosse 6
Fosse 7
OLD GERMAN FRONT LINE
BRITISH FRONT LINE
Loos Rd Rdt
Lens Rd Rdt
Chalk Pit Wood
Puits 14
Bois Hugo
Chalet Wood
Hill 70 Rdt
Annay 1¾m
Cité St. Auguste
LOOS
Loos Crassier
Garden City
Copse
Chalk Pit
Fosse 12
Dynamitière
Vendin le Viel 2m
Harnes 3m
Loison 1½m
Douai 10m
North Maroc
Grenay
South Maroc
Fosse 5
Double Crassier
Fosse 11
Cité St Edouard
Cité St Laurent
Cité St Pierre
LENS
Loos
KEY
••• English outposts on night of 25th Sept.
XI XI English line on morning of 26th c XI Corps jumping off point.
Divisional boundaries
Direction of maximum penetration on day of 25th.

cavalry. It was not until nightfall that they began to get clear of the complex of roads and lanes that were still crammed to capacity with supply and medical echelon and now awash under the heavy rain, and began to deploy in their final assembly positions behind Lone Tree Ridge.

As the men were formed up in the darkness for their cross-country march to the Lens–La Bassée road the confusion was considerable. Neither division had had longer than two weeks in France and their total training period in England had been no more than four months. They had, moreover, only a slight leavening of regular officers and N.C.O.s. None of the divisional staffs were familiar with the ground and there had been no time to issue large-scale maps. The men were soaked to the skin and, as the kitchens had been left behind, there was no hot food available. Thus it is not surprising to find that their deployment was only threequarters complete by day-break. By this time the troops had been continuously on the move for over eighteen hours, with only the customary halt of ten minutes in the hour while on the road. Even more serious was the fact that a thick mist, hanging over the area at dawn, led to the divisional artillery getting into positions some half-mile ahead of that allotted to them, so that when it cleared they were in full view of the German batteries about Haines and Hulluch. Thus the guns were effectually neutralized from the start.

But in spite of the complete exhaustion of all ranks, their morale was high. This was to be the division's first action, but they had been told that all that was required of them would be a long march in pursuit of a demoralized enemy. Both Haig and Haking, the corps commander, had assured them that they would 'not be put in unless and until the Germans are completely smashed and retiring in disorder'.

However, in contrast with his general assurance, Haig's *explicit* orders for the 26th were that the divisions were to continue the battle. These orders, issued from 1st Army H.Q. at 11.30 on the night of the 25th, while the wretched troops of the 21st and 24th Divisions were still stumbling about in the darkness on Lone Tree Ridge preparing for their cross-country march, emphasized the importance of the various 'attacks' that were to start in at 11 a.m. the following day.

At 9 a.m. Haig saw the three corps commanders concerned at his headquarters. He had breakfasted. He enlarged on the orders of the previous evening. The divisions were to push on through the German 'Second Position' and take the Haute Deule Canal, some five miles distant, as their objective. Such an advance, he said, might turn the whole enemy position to the south and even force the Germans to evacuate Lens itself.

If given in good faith this appreciation must be reckoned ludicrously optimistic. Nor did it square well with his assurances that the battle had entered a pursuit phase. No blame for Haig's mental confusion can be attached to Intelligence, who had accurately predicted the German dispositions. They had also sent a warning to 1st Army H.Q. that the German local reserves (a total of five divisions in the threatened sector) could begin to reach the battlefield within twelve hours of the alarm. In point of fact twenty-two additional battalions arrived in the battle area within twenty-four hours, so that by zero-hour on the 26th the German 'Second Position' was as strongly held as had been the front line at the time of the original assault the day before.

In contrast to the attack of the previous day—which had, at least, been preceded by a four-day artillery bombardment and a half-hour discharge of gas along the entire front, and had, moreover, been carried out by four selected divisions trained for the assault in every detail for weeks beforehand—the hapless 21st and 24th Divisions were expected to cross No-Man's-Land in broad daylight with no gas or smoke cloud to cover them, with no artillery support below divisional level, and attack a position as strongly manned as had been the front defences and protected by a formidable and intact barbed-wire entanglement.

. . . . .

The German 'Second Position' was in the form of a tall, shallow 'D'. Originally, in depth, it had consisted of an 'ID' with the village of Loos lying between the two letters, and the 'I' representing the first line. But this 'I' had been overrun in the previous day's fighting. The very strong enemy 'Second Position' followed the curve of the 'D', swinging round from the twin villages of Hulluch and Benifontaine in the north down to the Hill 70 Redoubt in the south, and giving terrible enfilade and cross-fire over the gently sloping Loos Valley. The bar of the 'D' was formed by the Lens road that stretched straight down from north to south, white with chalk dust, marked by an occasional leafless tree. Inside this area was a mass of small quarries and mineshafts, and a densely wooded copse, the Bois Hugo. These, as has been seen, were evacuated by the Germans during the fighting of the 25th but were reoccupied as reinforcements came up, and formed an excellent outpost line for machine-gunners and sharp-shooters. The diary of the 15th Reserve Regiment records:

'One battalion in particular had an excellent position along the edge of a disused quarry overgrown with thick bushes and scrub. They were well concealed from view, and yet had a perfect field of fire to front or flank. Four machine-guns were placed in position there, with the champion machine-gunner of the regiment at one of them.'

A tragic aspect of the situation lies in the fact that all of this area, and large stretches of the German 'Second Position', could have been taken for the picking the previous afternoon. A battalion of the German 26th Regiment, marching up from Annay to occupy the sector between Hulluch and the Bois Hugo, was told to make all haste as the British might already have entered it. So probable did this in fact seem that the battalion deployed half a mile from it just before dusk and advanced against it in extended order, only to find it empty. During the night many other opportunities had passed; two more battalions, of the 153rd Regiment, came up into the Bois Hugo sector of the second-line position south of the 26th Regiment and advanced into the wood itself at dawn, under cover of mist. As this cleared they attacked the outposts of the 63rd Brigade at the western end of the wood and drove them back, thereby enfilading the front position of the brigade immediately north of the wood and forcing them to retire back across the Lens road. As the morning wore on the Germans in the wood were reinforced by a battalion each of the 93rd and 165th Regiments and extended their line southwards, joining up with the reinforcements—a further six battalions strong—that had arrived to strengthen the position at Hill 70.

These, and other minor local counter-attacks carried out by the Germans with the intention of improving their defensive position, must surely have given 1st Army H.Q., and to both Haig and Haking, ample warning that an unprepared attack by two untrained divisions was unlikely to succeed. But the question of revising the order in the light of the Intelligence reports does not seem to have been considered. And so the stage was set for a repetition—at a distance of sixty-one years, in slow time, under conditions of infinite squalor and magnified in scale a hundredfold—of the charge at Balaclava. For the set-piece attack of 11th Corps, that was to be launched in the broad light of an hour before noon on the 26th, was as futile, and as foredoomed, as that of the Light Brigade.

As the morning wore on the British perfected their order of battle. Theirs was a depressing situation. They had had to cross the No-Man's-Land of the previous day, that was littered with the corpses of the Devon and Highland Regiments, lying in long straggling rows as the German machine-guns had traversed along their ranks. Among these were still many wounded who called out piteously to the newcomers for water and assistance. As assembly points the two divisions were using the former German front line. To make access to this easier, engineers had cleared gaps in the wire at regular intervals, but no one had yet had time to remove the contorted, lifeless figures that still hung at so many points on the entanglement. In the trench itself, and in the adjacent dug-outs, were pockets of gas, and many German dead, hideously yellow and blue in colour. The stench was frightful.

But the spirit of the men was unshaken. The official historian records that: '. . . they were delighted at the prospect of getting at the enemy after the exertions and frustrations of the last few days', although they had had hardly any food, and no sleep for forty-eight hours.

Just after 10 a.m. a desultory pattern of artillery fire, unworthy of the term 'bombardment', was thrown at the German positions. Without their own artillery the 21st and 24th were paying the price of Haig's 'flexible' distribution of guns for the first stage. Indeed the gunners could have no very clear idea of where the German emplacements were located and simply fired off patches of shells, assorted H.E. and shrapnel, at likely looking points. The Germans suffered no casualties and the wire remained intact. This fire lasted some twenty minutes, and then for half an hour the front was practically silent.

. . . . .

Punctually at eleven o'clock the British rose out of the ground. Peering across the shelving valley of rank grass, slag and white chalk craters, the German look-outs could see column after column moving up in close formation at the crest of Lone Tree Ridge, the officers on horseback, marshalling successive battalions as they rose out of the old German trenches and formed up in a dense mass.

At first the effect was unnerving. Not since the German attacks in the closing days of the first battle of Ypres had such dense masses of infantry deployed for a daylight assault. Sheer weight of numbers must, it seemed, carry the British through the thinly spread German outposts. The colonel of the 15th Reserve Regiment has described how he was walking in the main street of Hulluch when an experienced N.C.O. from the Machine Gun Company came running up to him and shouted out, 'Two divisions . . . we will be surrounded . . . we must retire. . . .' A number of men were following close behind, panic-stricken. But almost simultaneously another officer, who had been watching the situation from a housetop, came up and told him that the situation was not so serious. 'The machine-gun and rifle fire from our position is terrific and no enemy can possibly advance across the open against it.' Quickly, extra detachments were organized and sent into position. For fully ten minutes the Germans held their fire as the two divisions deployed in column of extended line and started obediently off on their progress down the gentle slope towards the Lens road. It was a tense moment for the enemy, watching in silence until, as the leading columns of the 24th Division passed under the south-east front of Hulluch, at a range of 1,000 yards, the order to fire was given.

The diary of the 15th Reserve Regiment records that:

> 'Ten columns of extended line could clearly be distinguished, each one estimated at more than a thousand men, and offering such a target as had never been seen before, or even thought possible. Never had the machine-gunners such straightforward work to do nor done it so effectively. They traversed to and fro along the enemy's ranks unceasingly. The men stood on the fire-steps, some even on the parapets, and fired triumphantly [*jauchsend*] into the mass of men advancing across the open grass-land. As the entire field of fire was covered with the enemy's infantry the effect was devastating and they could be seen falling literally in hundreds.'

As the British infantry advanced they started to come across little pockets of dead and dying from the detachments of the 2nd Brigade that had pressed too far the previous day.

Some of these, delirious, stood up and screamed at them to turn back, or to fetch stretcher-bearers, or to duck down and join them in an adjacent crater. But the discipline of the two divisions never wavered. Slightly in front of the lines walked the subalterns, shouting encouragement: 'Come on, me lads, we're nearly there.' 'It won't last long.' 'We'll soon be at 'em.' 'Show 'em what we are,' and so forth.[1]

And indeed the German diary noted with amazement:

'In spite of it [the intensity of the fire] the extended columns continued their advance in good order and without interruption. When they reached the Lens road one of our companies advanced from the Hulluch trench in an attempt to divert the attack, but only a small party of the enemy swung round to meet it, the mass took no notice and went on regardless past the southern front [of the village]. Here they came under the enfilade fire both of the troops lining the position and of a battery of artillery concealed in the village. Their losses mounted up rapidly and under this terrific punishment the lines began to get more and more confused. Nevertheless they went on doggedly right up to the wire entanglement.'

This barrier consisted of hard steel barbed wire, too thick to be cut with the hand-clippers that had been issued to some sections, braced and criss-crossed among pine stakes and pit-props driven thirty-five centimetres into the earth. Its height was over four feet and its depth across five metres, or nearly nineteen feet.

Desperate, the men hurled themselves at it in frenzy; some tried to scramble over it as one might a thick yew hedge, others pulled at it with their bare hands; still more ran up and down along its edge in the hopes of finding a gap that might have been cut by shellfire, until they were cut down.

1. Corporal J. Woosnam of the 8th East Yorks of 62nd Brigade of 21st Division has told me: 'The Lieutenant leading our Company, Harris or Harrison I think he was called, kept on talking all the time that we were going forward. He said the same thing over and over again—"Come on, my lads, show them what we are." After we had been advancing for about ten minutes he was hit by a burst of machine-gun fire in the stomach which lifted him right off the ground. He was calling for water and we gave him some although he was going to die, which he did in a few minutes' time.'

The German diary continues: 'Confronted by this hopeless impenetrable obstacle and faced by continuous machine-gun and rifle fire the survivors began to turn and retire in confusion, though scarcely one in ten that had come forward seemed to go back again.'

On the right flank the 21st Division was being dashed to pieces in like manner. The diary of the 153rd Regiment tells the same sort of story as that of the 15th:

'. . . dense masses of the enemy, line after line, appeared over the ridge, some of their officers even mounted on horse-back and advancing as if carrying out a field-day drill in peacetime. Our artillery and machine-guns riddled their ranks as they came on. As they crossed the northern front of the Bois Hugo, the machine-guns there caught them in the flank and whole battalions were annihilated. The English made five consecutive efforts to press on past the wood and reach the second-line position, but finally, weakened by their terrible losses, they had to give in.'

For the troops attacking the western end of the wood and struggling up the bare slopes of Hill 70 conditions were, if anything, worse.

One of the German battalion commanders spoke later of the revolting and nauseating impression made on them all as they watched the slaughter; so much so that after the retreat had begun they ceased fire. Before them was the 'Leichenfeld [field of corpses] von Loos', and, as among them dozens of khaki-clad forms rose up once again and began to limp and crawl back to their own lines, 'no shot was fired at them from the German trenches for the rest of the day, so great was the feeling of compassion and mercy for the enemy after such a victory'.

There had been twelve battalions making the attack, a strength of just under ten thousand, and in the three and a half hours of the actual battle their casualties were 385 officers and 7,861 men. The Germans suffered no casualties at all.

In the late afternoon, as the remnants of the two divisions rallied once more on Lone Tree Ridge, General Haking, the corps commander, came down from his H.Q. and moved among them, asking 'What went wrong?' The answer that he got from all, according to the *Official History*, was 'We did not know what it was like. *We will do all right next time.*' . . . Such was the spirit of those who had answered Kitchener's call 'Your Country needs YOU.'

For these men were volunteers. They were the flower of the richest, most powerful, nation on earth. Behind them stretched the ordered childhoods of Victorian Britain; decency, regularity, a Christian upbringing, a concept of chivalry; over-riding faith in the inevitable triumph of right over wrong; such notions were imbued in them. This had been their first time in action, but if these were the rules of the game, well then, they would conform.

13

The Dismissal of Sir John French

A very small memento, my dear, dear Douglas, of our long and tried friendship proved 'in sunshine and in shadow'.—J.F.

Inscription on a flask of Haig's,
now at Bemersyde

BACK in London the air was thick with rumour. After dusk, as the October mists closed on the city, the hospital trains would draw into Charing Cross and discharge their groaning cargo. Many of the wounded talked of the impossibility of the tasks that they had been given, of the hopeless sacrifice of their comrades. It was not difficult to equate the casualty lists and the miserable stretches of German line gained that, even on maps of the largest scale, looked thin and insubstantial. In his diary French noted petulantly:

'There was another military debate in the House of Lords on Monday night. The general tone adopted was to belittle and "crab" all the work which has been done by the Army in France. Lord Milner said, in so many words, that the battles of Neuve Chapelle and Loos instead of victories were in reality "defeats". Lord Courteney adopted the same line of argument.'[1]

Both he and Haig still clung, though without any real conviction now, to the hope that the German line in the Loos sector might yet be broken before winter set in. The thought of another winter in the trenches, which each had

1. French, p. 320.

felt during their periodic bouts of optimism might no longer be a necessity, was disquieting. Not only was supply of many essentials inadequate in quality and distribution, but the unsuitable terrain and situation of the troops in many sectors made it necessary to consider withdrawals to more favourable ground if their discomfort was not to be too acute.

Haig was opposed to this notion. By now his relations with French were very poor although no open breach had taken place. On the 2nd October he wrote: 'It seems impossible to discuss military problems with an unreasoning brain [French] of this kind. At any rate no good result is to be expected from so doing.'

As Haig would not agree to any withdrawal, French concentrated on persuading his ally to take over stretches of the British line. He painted the prospects of immediate glory in glowing colours:

'Foch came to see me this morning. I told him that if I was in the position of C.-in-C. of the whole western Allied front I would put every available man in just north of Hill 70 and "rush" a gap in the enemy's line. I should feel quite confident of success.'[1]

But Foch had a habit of scrutinizing the projects of colleagues more critically than he did his own. Tactfully he replied that: 'It would be very difficult to organize so big an attack in so small an area.'[2]

In point of fact the British troops were, during the first week in October, being subjected to a series of determined and effective counter-attacks that were steadily prising from them all the ground that had been gained at such cost on the 25th September. Haig noted that: 'On 2nd and 3rd October the 28th Division steadily lost ground so that we no longer had a line of trench suitable for launching an attack.'

He added the by now customary, though scarcely relevant,

1. French, p. 371.
2. Liddell Hart, *Foch*, p. 202.

comment that: '. . . The fact is that Sir John seems incapable of realizing the nature of the fighting that has been going on and the difficulties of getting fresh troops and stores forward and adequate communication trenches dug.'

On the Friday, 8th October, Haig drove over to French's headquarters at St. Omer and found him in 'chastened' mood. The first rumblings of protest from London at the massacre of the 26th September were beginning to be audible.

'Sir John read me a letter which he had received from Lord Kitchener asking for a report on the action of the 21st and 24th Divisions. Some of the wounded had gone home and said that they had been given impossible tasks to accomplish and that they had not been fed.'

Enquiries at this level, particularly when the actual results of operations were so meagre, boded no good for the commanders. And, sure enough, the following day, Saturday, Haldane turned up from England with only the barest notice and asked himself to lunch at Haig's headquarters. After the meal he and Haig repaired to the latter's room and the questions started. Haldane excused himself by saying that feelings were 'very strong' on the subject in England and that he had come to France 'to help in arriving at the truth'.

If Haig was disconcerted by this he quickly recovered his posture, and submitted an impromptu report[1] that was shamelessly critical of his superior. He ended by saying that 'the arrangements for the supreme command are not satisfactory. . . . Many of us feel that if these conditions continue it will be difficult ever to win.'

However it was not for nothing that Haldane had been thirty years in politics and he returned to England with an open mind.

The following day Robertson, French's Chief of Staff, travelled to London. As has already been explained Robertson enjoyed the worst relations with French. But his

1. Reproduced in Appendix 5.

friendship with Haig was of long standing, and was fortified by ties of mutual dependence which both hoped to strengthen in the immediate future. Although there is no direct evidence for this it seems reasonable to assume that Robertson consulted with Haig before the journey. On arrival in London he made contact with Lord Stamfordham, the King's private secretary, and the question of 'replacing' Sir John French was raised. An audience was arranged and at it the King listened attentively to French's shortcomings as catalogued by his Chief of Staff. No record is available of this meeting, but its substance must have been favourable to Haig, for at its conclusion Robertson journeyed post-haste back to 1st Army H.Q. Here Haig told him that:

'Up to date I have been more than loyal to French and done my best to stop all criticism of him or his methods. Now, at last, in view of what happened in the recent battle over the reserves . . . and of the seriousness of the general military situation, I have come to the conclusion that it is not fair to the Empire to retain French in command on this, the main battle front. Moreover none of my officers commanding Corps have a high opinion of Sir John's military ability. In fact they have no confidence in him.'

Haig went on to make the surprising claim that he got on better with his Allies:

'French does not get on with the French. Joffre seems to have no great opinion of his military views and does not really consult with him. It is most important at the present time to have someone to put the British case and co-operate with the French in aiming at getting decisive results in their theatre of operations.'

All this was duly communicated to the King. His Majesty decided, before taking any further steps, to come to France and see things for himself. Three days later he arrived at Boulogne where he found an uneasy French waiting on the quay. Owing to the shortness of notice received, there had been no time to prepare royal quarters and the King had to live in his train while these were being got ready for him at Aire. French, sensing perhaps a coolness and warned by his friends in England of impending trouble, took advantage of the dislocation of the royal schedule to slip back to England for a few days.

But before Sir John got back, on 24th October, the royal Mess was installed and Haig had been invited to dinner. After the meal he returned once again to the subject:

'I told him [the King] that the time to have removed French was after the retreat, because he had so mismanaged matters. . . . Since then, during the trench warfare, the Army had grown larger and I thought at first that there was no great scope for French to go wrong. I have therefore done my utmost to stop criticisms and make matters run smoothly. But French's handling of . . . the last battle, his obstinacy and conceit, showed his incapacity, and . . . I therefore thought strongly that, for the sake of the Empire, French ought to be removed.'

He added, superfluously it may be thought, that 'I, personally, was ready to do my duty in any capacity . . .'

The King listened attentively. That afternoon, he confided to Haig, he had had an informal chat with General Haking who had told him much the same thing. General Robertson, too, was very critical of Sir John. When Haig got back to his headquarters at Hinges at midnight, he must have felt that he was on the point of achieving a vital personal victory.

But then, two days later, with things still in the balance, there was an unfortunate occurrence. The King was thrown by a mare. He was badly bruised. He had to take to his bed for several days.

The incident was the talk of the Army and dominated, as well, the pages of the newspapers at home. Speculation and gossip about changes in the High Command were forgotten. The smooth flow of events which must, it had seemed, carry Sir John from his position was interrupted. For not only was one of his foremost opponents incapacitated, but it would have been most unseemly for any note of discord to have intruded on the harmonious messages of sympathy that poured in from all sides. The *status quo* must be preserved at present, and later on, as the first sprinkling of snow covered the Leichenfeld von Loos, the events of the 26th September would already be half forgotten.

For General Haig the incident had a particularly trying personal aspect; for it was his mare that the King had been riding. It had been his personal responsibility to see that she was exercised and 'quiet'.

What a compliment it had been that His Majesty should have selected one of Haig's personal stable! And how nightmarishly things had turned out! For if he claimed that the mare had been thoroughly exercised it was a reflection on the King's horsemanship, but if he admitted that his staff had been negligent in quieting her beforehand . . . Haig's diary devotes much space to an account of, and apologia for, this incident; more space, indeed, than is occupied by the whole of the battle of Loos.[1] On and on rambles the text, as we read that it was a chestnut mare, that Haig had ridden her regularly for over a year, that she had been tried the day before with cheering men and people waving flags, that hats—not flags—were waved, that the grass was wet, that the ground was slippery, that the King seemed to clutch the reins *very firmly* (a tricky passage, this), and to pull the mare backwards, that the cheering would have upset any horse at such a distance, and so on.

That evening Sir Derek Keppel telephoned to Haig and told him that the King was to remain in bed for a couple of days and that he could not receive visitors. He much re-

1. I am referring to the *published* text of Haig's diary. Robert Blake has kindly pointed out to me that in the original document (which has not been published) this ratio does not apply.—A.C.

gretted that he would not be able to see Haig again before he returned to England. The diary records also:

'He was desired by the King to say that His Majesty knew very well that the mare had never done such a thing before and that I was not to feel perturbed at what had happened (or words to that effect).'

All the same, the result of the whole episode must have seemed unsatisfactory to Haig.

Sir John quickly sensed the shift in fortune and made every effort to consolidate his position. As soon as the King had left France, he, too, journeyed to London where he consulted with his friends—themselves a body of some influence, and including Colonel Repington, the military correspondent of *The Times*—and in concert with them he evolved further measures to protect his interests.

While he was in the capital he attended two Cabinet meetings. He found that body ineffectual and pusillanimous and reported this—not without a glint of triumph, one may suspect—to Haig on his return.

'Everyone seemed desirous to speak at the same moment. One would say "Please allow me to finish what I am saying." Another would interrupt, and a third would shout from the far end of the room that he meant to have his say on the matter too. Poor Lord Crewe[1] feebly rapped the table and said "Order please!" in a disconsolate sort of way. . . .'[2]

. . . Not much chance of *that* Cabinet dismissing the Commander-in-Chief—and small wonder that Haig found 'Sir John seemed in excellent spirits since his return. . . .'

As it turned out, however, the measures initiated by French in London proved his undoing. On the 2nd November his 'full' despatch dealing with the attack on the 25th September appeared in *The Times*—filled with needless inaccuracies that could with facility be exposed by anyone

1. Crewe was presiding during Asquith's illness.
2. Haig, *Diaries*, p. 111.

who had access to the relevant documents, telegrams, orders, etc. Haig, of course, was just such a person, and any reservations that he might have felt about so flagrant a breach of the canons of loyalty and discipline must have been finally dispelled by the text of an article by Colonel Repington, published alongside French's despatch. In this article the military correspondent of *The Times* regretted that the operations at Loos had not been under the direct command of Sir John, whose abilities were by implication contrasted favourably with those of Haig.

Accordingly two days later a long letter went out from 1st Army H.Q. to French's staff pointing out these 'misstatements' and, in particular, enclosing copies of telegrams showing that the 21st and 24th Divisions were not placed under Haig's orders at the beginning of the battle, and that the Guards Division was not placed under his orders until 4.15 p.m. on the 26th. (French had stated that it came under Haig's command 'that morning'.)

If anything were needed to illustrate the very secondary mental calibre of Sir John French, and his impaired capacity for anticipation, it is his conduct of 'The affair of the Reserves at Loos'. He allowed himself to be completely outmanœuvred—and on a mere technicality. Haig had selected as his grounds of dispute the fact that the reserves had not been brought forward *in time*. The fact that this, in itself, constituted an oblique admission that he had committed them to battle when he knew that it was too late, and thus subjected them to a pointless slaughter, seems not to have occurred to Haig. And a rejoinder was obviously denied to Sir John.

When Haig's letter was received, asking 'that these facts should be put on record', French collapsed and retired to his room.

He remained in bed—'staying out' in the parlance of Etonians—for thirty-six hours. Finally he pulled himself together sufficiently to draft a letter stating that 'this correspondence shall henceforth cease'. And that '. . . the statements in question are substantially correct, and call for no amendment'.

After a further three days of delay during which interest and indignation, as reflected in the columns of *The Times*, rose to a high pitch, French sent for Haig. He told him that all the correspondence would be sent to the War Office together with a covering letter, the text of which he would first show to Haig. He also disclaimed any connection with Repington's article.

But it was too late. Haig brushed aside French's apology and declined to co-operate in hushing the matter up on the grounds that '. . . my duties as G.O.C. First Army take up all of my time'. When Lord Esher turned up at his head-quarters the following day Haig, with a subtlety that he was more prone to display in the furtherance of his personal ambition than in the planning and execution of his military campaigns, hardly mentioned French but concentrated instead on running down another of his military superiors, Kitchener. At this time the Cabinet were pre-occupied with the importance of ridding themselves of 'K'. It may be suggested that one of Sir John French's few commendable traits, in the eyes of the politicians in England, lay in his implacable hostility to the Secretary of State for War, and Haig probably saw that the Cabinet would have to be sure in their minds that these views were shared by any successor to the Commander-in-Chief that they might nominate.

'Lord Esher asked what should be done with Lord Kitchener. I replied, "Appoint him Viceroy of India." Trouble is brewing there and in Burma, and some blood-letting will become necessary for the health of the body politic!'

There is no doubt that other reasons, also, prompted Haig to get rid of Kitchener besides his desire to align himself with the Cabinet. He was apprehensive of Kitchener's strategic vision, that dwarfed all other contemporary minds save those of Churchill and Lloyd George; of his autocratic manner; and of the devotion that he commanded from the public at home:

'. . . in my opinion it is important to remove Lord Kitchener from the Mediterranean and Egypt because wherever he is, by his masterful action he will give that sphere of of the operations an undue prominence in the strategical picture.'

There was also the consideration that Kitchener stood in the way of an appointment that Haig was determined to secure, namely that of Robertson to Chief of the Imperial General Staff—or at any rate in the way of a full and free exercise of the powers that went with such a post. In the course of Lord Esher's visit Haig pressed vigorously for Robertson's appointment and also put forward a variety of other proposals for clipping Kitchener's wings.

Robertson, in the meantime, was in London. On the 15th November he wrote:

'My dear Haig, I have got your letter. Many thanks[1]—I shall try my best to help put things straight. . . .'

And on the 17th: '. . . just a line to say that I am doing my best. . . .'

And on the 20th: '. . . it would do the world of good if you could come over yourself for a few days. . . .'

Haig needed no further encouragement. He left France the day after receiving Robertson's call and, on the 23rd November:

'. . . Doris and I went shopping. At 1.45 we lunched at 10, Downing Street. I sat next to Mrs. Asquith, and after the ladies left the room I sat by the Prime Minister. . . .'

There seems little doubt that at this meeting Haig was briefed for the Cabinet meeting on the following day. At this he urged further and more elaborate measures for curtailing

1. Presumably for Haig's 'work' with Lord Esher.

Kitchener's power and influence.[1] These were, naturally, welcome, and as Bonar Law drove Haig home in his car '. . . he stated how Lord K had misled the Government and wondered what appointment could be found for him to remove him from London!'

In the light of these intrigues there is a touch of irony about Haig's note of his next meeting with Kitchener:

'Lord K. was most friendly. . . . He said that today [3rd December] he had written to the Prime Minister recommending that I should be appointed to succeed to Sir J. French. If the P.M. did not settle the matter today he would again press for a settlement tomorrow, but in any case he had taken the matter in hand and I must not trouble my head over it. Meantime, he said that I must not be afraid to criticize any of his actions which I found unsatisfactory; he had only one thought, viz, to do his best to end the war. . . . He again kindly told me that he would look after my interests and wished me good luck.'

Now Haig had on his side the King, the Cabinet, and 'K'. It was impossible for French to resist any longer, and, on December 10th, Haig wrote:

'About 7 pm I received a letter from the Prime Minister marked "Secret" and enclosed in *three* envelopes!

'It was dated 10 Downing Street, Dec: 8th 1915, and ran as follows:

' "Sir John French has placed in my hands his resignation of the office of Commander-in-Chief of the Forces in France. Subject to the King's approval I have the pleasure of proposing to you that you should be his successor. I am satisfied that this is the best choice that could be made in the interests of the Army and the Country." '

There were one or two final touches to go on the canvas before the picture was complete. As his successor, Haig recommended, for G.O.C. 1st Army, Sir Henry Rawlinson,

1. Appendix 6.

noting at the same time, 'Though not a sincere man he has brains and experience.'[1]

And then there was the last meeting with Sir John. It could not be pleasant. In alarm Haig noticed that his erstwhile superior was 'not looking very well, and seemed short of breath at times'. In still greater alarm he followed Sir John over to a remote part of the compound in obedience to the latter's request that there was a 'delicate personal matter' that he wished to speak about. Was there to be, at the last, some undignified personal scene? Some bitter recrimination that might end in shouting, in gesture, that would be noticed by those standing around them?

But no, even this slight price was not to be paid. All that Sir John wished to say was that he had wanted to give Winston Churchill[2] an infantry brigade, which idea had been vetoed, but that he was none the less anxious that he should have a battalion. Haig replied that he had no objection, and the interview was at an end.

Now, at last, Haig had reached the summit. His was the command of the greatest army that the Empire had ever put in the field in the past, or was ever to amass in the future. A body whose heroism and devotion was such that they could twice in two successive years be ravaged in hopeless offensives, who were in a single day to lose more men than any other army in the history of the world, whom, after twenty-seven months of slaughter and exhaustion, he was to leave so perilously exposed that they were nearly annihilated—and yet whose fortitude was such that they could still, after three years, be brought to final victory.

The change in Command became official on 1st January 1916. The evening before there was

'A regular New Year's beano, more like a London New Year's Eve festival than a war one. There were present the

1. Rawlinson was not in fact appointed. Instead the job went to Monro, whose obesity had earlier aroused Haig's comment. See p. 24 above.

2. At this time Winston Churchill, going through a period of reaction after the Dardanelles, was doing a stint in the front line.

Duke of Teck, on D.H.'s right, General Macready, Sir A. Sloggett, General Butler, Alan Fletcher, Sir Philip Sassoon, etc. I do not think that any of us spoke about the present war all through dinner. . . . Sloggett was the life and soul of the party with his yarns, some of which were libellous and few of which would have passed muster in a drawing-room. D.H. never shines at a dinner, but he was obviously in very good spirits and kept silence merrily.'[1]

But in the line the trenches were already heavy with mud and that Christmas on the leave platforms at Victoria there began to be heard the chorus of a new song:

> I don't want to die,
> I want to go home.
> I don't want to go to the trenches no more,
> Where the whizz-bangs and shells do whistle and roar.
> I don't want to go over the sea,
> To where the alleyman will shoot at me,
> I want to go home
> I don't want to die.

1. Charteris, *G.H.Q.*, p. 129.

Appendix 1

The Charge of the 9th Lancers at Frétoy, on the morning of 7th September 1914, from *1914*, by Field Marshal Viscount French of Ypres:

'On reaching Frétoy the village of Moncel was found to be occupied by a patrol of Germans, and was taken at a gallop by the leading troop, followed by the one remaining machine-gun of the regiment. About a troop and a half, accompanied by the Commanding Officer, Lieutenant-Colonel D. Campbell and Major Beale-Brown, moved up on the left of the village. Shortly afterwards two squadrons of the *1st Garde Dragoner* charged the village and drove out the troop of the 9th Lancers after a little street fighting. A third *Dragoner* squadron then came up to the village from the north in support. The troop and a half of the 9th Lancers, led by the Commanding Officer and 2nd in Command, attacked this squadron in perfect order, charged the left half of the German squadron and pierced it with loss, both sides facing the charge; the Germans at a 15-mile rate and the 9th Lancers at speed.

'In this charge Lieut.-Col. Campbell was wounded in the arm by a lance and in the leg by a bullet, both wounds, however, being slight. The adjutant, Captain G. F. Reynolds, was severely wounded in the shoulder by a lance. Lieutenant Alfrey . . . was killed while extracting the lance from Captain Reynolds. Our casualties were slight, one officer and two men killed, two officers and five men wounded. The number of Germans left on the ground was considerable.'

This was Lieutenant-Colonel, later General, David Campbell, who had ridden his famous gelding, 'Soarer', in the Grand National of 1912. He commanded the 21st Division in the opening weeks of the battle of the Somme.

Appendix 2

WARNINGS RECEIVED BY INTELLIGENCE OF IMPENDING GERMAN GAS ATTACK

Bulletin of the French 10th Army, 30th March 1915

Employment of Asphyxiating Gases by the Germans

'According to prisoners of the XVth Corps, there is a large supply along the whole front in the neighbourhood of Zillebeke (this was where the enemy first intended to use gas) of iron cylinders, 1·4 metres long which are stored a little in rear of the trenches in bomb-proof shelters or even buried. They contain a gas which is intended to render the enemy unconscious or to asphyxiate him. It has not yet been made use of, but the pioneers have received instructions regarding its employment; the cylinder is laid on the ground pointing towards the enemy and is opened by withdrawing the cap; the gas is forced out by its own pressure, and remains near the surface of the ground. In order that the operation may be without danger for the operator a favourable wind is necessary. The pioneer detailed to open the cylinder has a special apparatus attached to his head. All the men are supplied with a cloth pad to be placed over the nostrils to prevent the gas being breathed in. The inventor has been promoted lieutenant.'

2nd Army Report to G.H.Q., dated 15th April

'A reliable agent in the Detachment of the French Army in Belgium reports that . . . The Germans intend making use of tubes with asphyxiating gas, placed in batteries of 20 tubes for every 40 metres along the front of the XXVIth Corps (then, as far as was known, in the line on either side of Langemarck, wholly opposite the French). This prisoner had in his possession a small sack filled with a kind of gauze or cotton waste (cotton waste in a gauze bag) which would be dipped in some solution to counteract the effect of this gas.

'The German morale is said to have much improved lately, owing to the men having been told that there is not much in front of them.

'It is possible that the attack may be postponed, if the wind is not favourable, so as to ensure that the gases blow over our trenches.'

Bulletin de Renseignements sur le Détachement de l'Armée de Belgique, dated 16th April

'. . . the Germans have manufactured in Ghent 20,000 mouth protectors of tulle, which the men will carry in a waterproof bag 10 cm. by 17·5 cm. These mouth protectors, when soaked with a suitable liquid, serve to protect the men against the effects of asphyxiating gas.'

One further point that should be mentioned is that in their communiqué of 17th April the Germans said: 'Yesterday east of Ypres the British employed shells and bombs with asphyxiating gas.' To anyone familiar with the German mentality this should have been sufficient warning that they intended to do something of the same kind themselves and were putting the blame on their opponents in advance.

Appendix 3

THE SMITH-DORRIEN CORRESPONDENCE

'Advanced Headquarters, 2nd Army,
27th April 1915

'My dear Robertson,

'In order to put the situation before the Commander-in-Chief, I propose to enter into a certain amount of detail.

'You will remember that I told Colonel Montgomery (H. M. de F., General Staff, G.H.Q.) the night before last, after seeing General Putz's orders, that as he was only putting in a small proportion of his troops (and those at different points) to the actual attack, I did not anticipate any great results.

'I enclose you on a separate paper the description of the line the troops are on at this moment. I saw General Putz last night about to-day's operations, and he told me he intended to resume the offensive with very great vigour. I saw his orders, in which he claims to have captured Het Saas, but on my asking him what he meant he said the houses of that place which are to the west of the canal. He told me also that the success at Lizerne had been practically nil—in fact, that the Germans were still in possession of the village or were last night.

'From General Putz's orders for to-day, he is sending one brigade to cross the river east of Brielen to carry forward the troops on the east of the canal in the direction of Pilckem, and he assured me that this brigade was going to be pushed in with great vigour.

'It was not till afterwards that I noticed that, to form his own reserve, he is withdrawing two battalions from the east of the canal and another two battalions from the front line in the same part to be used as a reserve on that bank of the river, so the net result of his orders is to send over six fresh battalions to the fighting line and to withdraw four which had already been employed.

'I have lately received General Joppe's orders. He is the general commanding the attack towards Pilckem on the east of the canal, and I was horrified to see that he, instead of using the whole of this brigade across the canal for this offensive, is leaving one regiment back at Brielen, and only putting the other regiment across the canal to attack—so the net result of these latter orders with regard to the strength of the troops on the east of the canal for the fresh offensive is the addition of one battalion.

'I need hardly say that I at once represented the matter pretty strongly to General Putz, but I want the Chief to know this as I do not think he must expect that the French are going to do anything very great—in fact, although I have ordered the Lahore Division to co-operate when the French attack, at 1.15 p.m., I am pretty sure that our line to-night will not be in advance of where it is at the present moment.

'I fear the Lahore Division have had very heavy casualties, and so they tell me have the Northumbrians, and I am doubtful if it is worth losing any more men to regain this French ground unless the French do something really big.

'Now, if you look at the map, you will see that the line the French and ourselves are now on allows the Germans to approach so close with their guns that the area east of Ypres will be very difficult to hold, chiefly because the roads approaching it from the west are swept by shell fire, and were all yesterday, and are being to-day. Again, they are now able to shell this place, Poperinghe, and have done it for the last three days; all day yesterday at intervals there were shells close to my Report Centre and splinters of one struck the house opposite in the middle of the day, and splinters of another actually struck the house itself about midnight—in other words, they will soon render this place unhealthy.

'If the French are not going to make a big push the only line we can hold permanently and have a fair chance of keeping supplied would be that passing east of Wieltje and Potisje with a curved switch which is being prepared through Hooge to join onto our present line about a thousand yards east of Hill 60.

'This of course means the surrendering of a great deal of trench line, but any intermediate line short of that will be extremely difficult to hold, owing to the loss of the ridge east of Zonnebeke which any withdrawal must entail.

'I always have to contemplate the possibility of the Germans gaining ground east of Lizerne, and this, of course, would make the situation more impossible . . . in fact it all comes down to this, that unless the French do something really vigorous the situation might become such as to make it impossible for us to hold any line east of Ypres.

'It is very difficult to put a subject such as this in a letter without appearing pessimistic . . . I am not in the least, but as Army Commander I have to provide for every eventuality and I think it right to let the Chief know what is in my mind.

'More British troops could restore the situation but this I consider out of the question, as it would interfere with a big offensive elsewhere which is after all the crux of the situation and will do more to relieve this situation than anything else.

'Yours sincerely,

'H. L. Smith-Dorrien.'

Telephone message (record is in writing of C.G.S. Lieutenant-General Sir W. R. Robertson):

'2.15 p.m. 27th April

'C.G.S. to Second Army

'Chief does not regard situation nearly so unfavourable as your letter represents. He thinks you have abundance of troops and especially notes the large reserves you have. He wishes you to act vigorously with the full means available in co-operation with and assisting the French attack having due regard to his previous instructions that the combined attack should be simultaneous. The French possession of Lizerne and general situation on Canal seems to remove anxiety as to your left flank.

'Letter follows by Staff Officer.'

But in fact no letter was ever sent, the next communication being a telegraphed message sent, *en clair*, that same evening:

'Advanced Second Army, V Corps

'Chief directs you to hand over forthwith to General Plumer the command of all troops engaged in the present operations about Ypres.

'You should lend General Plumer your Brigadier-General General Staff and such other officers of the various branches of your staff as he may require. General Plumer should send all reports direct to G.H.Q. from which he will receive his orders. Acknowledge.

'R. Hutchinson, Major, G.S.'

Appendix 4

TRENCH WARFARE AT CLOSE QUARTERS

Extracts from the diary of Captain F. Hitchcock, of the Leinster Regiment, relating to conditions in the Ypres salient following Plumer's withdrawal to the new line in the summer of 1915:

'11th August

'The C.O. had a "pow-wow" with all the officers in the Battalion. He told us that we were ordered to consolidate the new position at Hooge. He said we would have to dig in and wire all night, and that we must be prepared for a counter-attack. His final orders were: "Go and tell your platoons what they are up against, and what to expect." At 7 p.m. we marched off for Hooge in battle order, each man carried sand-bags, and a pick or shovel. Algeo was in charge of the Company, as Caulfield had gone up in advance to take over the front line. Marsland was attached to Battalion Headquarters as Bombing Officer. We got to the front line at Hooge after a rough journey under shell-fire, over dead men and round countless shell-holes. At 11 p.m. we had taken over the Hooge sector and the mine crater from the last Battalion The Buffs . . . 16th Brigade. The order of the Battalion in the line was C and D Companies front line, with A and B in support and reserve. We had hardly taken over the line, when the Huns attacked our left flank, which was exposed. However, Algeo had posted the Company Bombers there, and with a handful of men armed with jam-tin bombs, succeeded in beating them back.

'By the light of the moon and the glow from the green-white star-shells one could just distinguish the serpentine course of the German lines running along the near side of the Bellewaarde or Château Wood, only 50 yards away. The leafless trees stood out in their shattered forms, and behind them was the lake reflecting the moonlight.

'My left flank was in the air, a barricade only separating us from the Germans. We actually shared our front line with the enemy! How this strange fact came about was as follows: the continuation of our front line running up the Bellewaarde ridge which had been captured by the 16th Brigade on the 9th had to be abandoned the same evening, as the trench was untenable owing to the enfilade fire which caused terrible havoc to the troops holding it. This enfilade fire came from the German positions on the high ground, on the extreme left flank at Bellewaarde Farm.

'The 16th Brigade, therefore, evacuated this enfiladed section and erected a strong, sand-bagged barrier in the trench with a good field of fire. After about twelve hours, the Huns cautiously worked their way along their old front line, from Bellewaarde Farm, and found that our troops had withdrawn. After the bombing attack had been successfully repulsed Algeo and his bombers flattened out the old German parapets, and filled in the trench in front of the barricade, so that the Huns could not approach this post under cover. Within 15 yards of our barricade, the enemy switched off their old front line into their old support line. Throughout the night, the enemy were very offensive with bombs and snipers. We did not retaliate, as we were too busily employed reversing the parapets, making fire steps and deepening the trench everywhere, as we were anticipating a bombardment and a counter-attack on the morrow. Serg. Bennett and his machine-gun section worked splendidly, and built two fine battle positions for their guns. All the men worked like Trojans on top of the parapets in their shirt-sleeves.

'The place reeked with the smell of decomposed bodies. They lay about in hundreds, on top of the parapets, in our trenches, in No-Man's-Land, and behind the parados. The British dead mostly belong to the 2nd York and Lancs, and the 2nd D.L.I. The dug-outs were full of dead Germans, those that were not, two only, were strengthened for occupation. While we were working bullets spat viciously all round, and we had several casualties.

'12th August

'Dawn broke at 4 a.m. and within half an hour I had two casualties. Pte. Bowes was killed by an explosive bullet in the head, and Pte. Duffey was wounded by an enfilade bullet from the Bellewaarde Farm. We buried Bowes in a disused trench behind our line. One could now make out the country all round perfectly, and what an appalling sight it was. Everywhere lay the dead. The ridge in our rear was covered with dead men who had been wiped out in the final assault of the German position: their faces were blackened and swollen from the three days' exposure to the August sun, and quite unrecognizable. Some of the bodies were badly dismembered: here and there a huddled-up heap of khaki on the brink of a shell-crater told of a direct hit. Haversacks, tangled heaps of webbed equipment, splintered rifles, and broken stretchers, lay scattered about. The ground was pitted with shell-holes of all sizes. A few solitary stakes and strands of barbed-wire was all that was left of the dense mass of German entanglements by our artillery. Several khaki figures were hanging on these few strands in hideous attitudes. In front of us, in No-Man's-Land, lay a line of our dead, and ahead of them on the German parapet lay a D.L.I. officer. They had advanced too far, and had got caught by a withering machine-gun fire from the Bellewaarde Wood. There was not a blade of grass to be seen in No-Man's-Land or on the ridge, the ground had been completely churned up by the shells, and any of the few patches of grass which had escaped had been burnt up by the liquid fire. Some 50 yards away, around the edge of the Bellewaarde Wood, ran the sand-bagged parapet of the German line on its serpentine course towards the shattered remains of Hooge.

'The wood itself had suffered severely from the shell-fire. Most of the trees were badly splintered, and some had been torn up by the roots. There was little foliage to be seen on any of the trees. All that was left of the once bushy-topped trees which lined the Menin Road were shattered stumps, and the telegraph poles stood drunkenly at all angles. Although numbers of the Durhams and the York and Lancs lay about in the open, yet our trench was full of German dead belonging to the Würtembergers.

'They lay in the dug-outs, where they had gone to seek refuge from our guns, in fours and in fives. Some had been killed by concussion, others had had their dug-outs blown in on top of them, and had suffocated. Our gunners had done their work admirably, and the strong cover made with railway lines and sleepers and with trunks of trees had collapsed under the fierce onslaught of our shells. The faces of the enemy dead, who had thus been caught or pinned down by the remnants and shattered timber of their death traps, wore agonized expressions.

'Here and there, where portions of the trench had been obliterated by the shells, legs and arms in the German field-grey uniform stuck out between piles of sand-bags. Thousands of rounds of fired and unexpended cartridges lay about the parapets, and ground into the bottom of the trench. German Mausers, equipment, helmets, and their peculiar skin-covered packs lay everywhere. The ground was littered with portions of the enemy uniforms saturated in blood. Serving in the Ypres salient one was not unaccustomed to seeing men blown to pieces and, therefore, I expected to see bad sights on a battle-field, but I had never anticipated such a dreadful and desolate sight as the Hooge presented, and I never saw anything like it again during my service at the front. The reason that Hooge was such a particularly bloody battle-field was due to the fact that it covered such a small area in the most easterly portion of the salient, and was not spread out over miles of open country like those battle-fields on the Somme in 1916. Hooge had been continually under shell-fire since the First Battle of Ypres in October, and the ridge which we had dug into had been captured and recaptured five times since April.

'At 5 a.m. some shells fell all along our line. Then all was silent and we realized the meaning of those dozen shells which traversed our line from left to right, ranging shots for a pukka bombardment. Within fifteen minutes of the burst of the last shot, a steady bombardment started all along our line.

'The enemy gunners carried out their work in a most systematic manner. They fired by a grouping system of five shells to a limited area, under 12 yards. Then they burst shrapnel over this area. This plan for shelling our position was undoubtedly successful, as three out of the five shells hit our trench, obliterating it, blowing in the parapet on top of the occupants, or exposing them to a deadly hail from shrapnel shells. Our casualties were beginning to mount up. A direct hit with a 5·9 knocked out six men of the Machine-Gun Section, Burlace, Cleary, and Scully being killed. As there was no communication trench the walking wounded "chanced their arms" going back over the ridge which was being raked by shrapnel fire, but the badly wounded had to lie in the bottom of the trench and wait until the cover of darkness to be carried back by the stretcher bearers. Some of these stretcher cases were, unfortunately, hit for a second time and killed.

'At 12.30 p.m. the shelling eased, and we got ready for a counter-attack. The order: "Pass along the word, fix bayonets," went along the line. We all, except the wounded who looked wistfully up at us, armed to the teeth, looked forward to Germans getting out of their trenches, but they did not. Although there were no wire entanglements of any description in front of us, as the single stretch of concertina wire had been cut by the first shells, yet we would have given hell to the Huns had they attacked. They obviously calculated on us retiring from our seemingly hopeless position, but we did not budge an inch. During the lull, the men dragged the wounded under better cover, dug out more funk-holes, and took the opportunity to "drum up their char". Shell-fire, the smell of powder, and the continual dust made us all very thirsty, and never did I relish a drink of tea more than that dixie-full which L/Corpl. Leanard and Pte. Coghlan shared with me. The dixie was chipped all round the edges and was blackened by smoke!

'At 3 p.m. exactly, the enemy started a second bombardment of our line. All along our trench they put down a terrific barrage of shells of every description. High explosives and crumps exploded on our parapets, leaving burning and smoking craters, and torn flesh, and above, screeching and whining shrapnel burst over us. We were shelled from all sides by guns of every calibre. We could not have been in a worse position, and it seemed that every enemy gun around the salient was turned on to our 400 yards of trench on the left of the Menin Road. Shells from the Bellewaarde direction enfiladed us, and blew in our few traverses: shells from the Hill 60 direction ploughed great rifts in our parados, and broke down our only protection from back-bursts, and now and then some horrible fragments of mortality were blown back from the ridge with lyddite wreaths.

'The whole place had become quite dark from the shells and

the clouds of earth which went spouting up to the sky. We could barely see twenty yards ahead throughout this terrible tornado of fire. Our casualties increased at such a rapid rate that we were all greatly alarmed, our trench had ceased to exist as such and the enemy shrapnel caused dreadful havoc amongst the practically exposed company. L/Corpl. Leanard, Privates Keenan, McKenna, Digan, and Shea of my platoon, had been hit, and Algeo got a direct hit on his platoon, killing 6032 Pte. Fay, and 3642 Pte. Lysaght, and wounding Privates Healey and Rattigan badly, and four of his N.C.O.s. If this went on much longer, the Boches would walk into our position without any opposition, as we would all be casualties. The shells came down with tantalizing regularity, which was nerve-racking.

'A most demoralizing effect is that of being smothered in sand-bags. Twice I emerged out of a heap of demolished sand-bags to find men hit on either side of me. It was extraordinary how one got to know and understand the men under shell-fire. " 'Tis different now beyont in Killyon, sir," said a man in my ear. "Ye gods, yes!" I replied. The man had seen me many times pre-war, as Killyon was only three miles from my home.

'I went up to the bend where Company Headquarters was situated. They had just got a direct hit and the stretcher bearers were on their knees bandaging some lifeless-looking forms. Another yell rang out for stretcher bearers from close to a smoking crater, and off Dooley ran to give first aid along the top of the trench into the blackness, and disappeared from view. Healy and Rattigan, who had been hit earlier in the day, lay alongside each other in the bottom of the trench. Algeo was standing beside Sergt. Bennett, who was sucking at an old clay pipe. Both wore an expression of defiance on their determined-looking faces. Rattigan was in a semi-conscious state, and blackened from head to foot with powder. Healey was in frightful pain: he had been badly hit in the stomach, and kept calling for water. "Mister Algeo, for the love of God give me a drink." "Stay quiet now, Healey, and you'll be all right soon." But he would not stay quiet. He then spotted me, and asked me for my water-bottle, but I could not give it to him. Reid came along and rinsed his mouth with water. "Can't ye keep quiet now, for a few minutes. Shure 'tis meself that will be bringing you along to the dressing station." But Healey would not stay quiet. "Holy Mary, Mother of God!"

Bang, crash! A shrapnel shell burst right over us, and Healy lay quiet for all time. He had been hit for the second time. " 'Tis as well, sir," said Morrissey. "He hadn't a hope: a piece of shell as big as your fist in his stomach!"

'The blackened bodies of our dead, and the badly wounded, lay about at the bottom of the trench, and it was impossible to move without treading on them. Every few minutes the call for the stretcher bearers would be heard. Then along came Morrissey with his first-aid bag, closely followed by Reid. "Steady, me lad," they'd say to a man who had lost his leg, but could still feel the toes of the lost limb tingling, " 'tis a grand cushy one you've got. Sure you're grand entirely, and when darkness sets in we'll carry you off to the dressing station, and then ye'll get your ticket for Blighty." How they stuck it, those company stretcher bearers, Morrissey, Reid, Dooley, and Neary. White men all!

'15th August

'The C.O. came over and told Caulfield that C and D Companies were to go back to the front line for two more days! He said that Piper should strengthen the barricade, and put out barbed wire: also that the positions had to be made as strong as possible, with traverses and flying traverses to stop the enfilade fire. He detailed me to bury all the dead at night.

'At 5.30 B Company got heavily shelled for over 1½ hours. The shrapnel-fire was terrific, and their line was completely enveloped in a dense mass of smoke from the bursting shells.

'We stood to in case of an attack, but the shell-fire did not materialize into anything. Daly's casualties were 7 killed and 21 wounded. These numbers included most of his N.C.O.s.

'When it was dark I set off with two platoons to bury the dead. It was a most unpleasant duty, as they were all men of the Durhams and York and Lancasters, who had been killed on the 9th in the charge. There were many other bodies lying out in this shell-churned area, and the ghastly stench of mangled corpses gripped us all by the throat as we carried out our task. It was very sad, but headless and armless got exactly the same treatment. We searched all for their identity discs, and their Army Books 64, and any other personal belongings for their next-of-kin. We salved their webbing equipment and rifles, and buried them in threes and fours in large graves. We buried some fourteen and returned to the reserve line, where we all got a rum issue. Barnett got a bullet through the stomach when he was guiding a working party of the 1st North Staffords along the Menin Road. Poole and Pearman were wounded and Louis Daly slightly, but he remained at duty, being the only officer left in B Company. Ducat, who was transport officer, returned to duty to assist him. C and D Companies took over the front line again at 9 p.m.

'16th August

'Barnett died of his wounds. The Doctor told us that he stuck his wounds splendidly, and that men who were only hit in the arms and legs were groaning all round him in the dressing-station. Barnett had a presentiment that he would get killed, and told us so when we got orders for Hooge. I relieved "Cherrie" Piper and Caulfield at 9 a.m. in the fire trench. The Brig. General came round to inspect the line with the C.O. The Brigadier said the Battalion had done splendidly, and that the place was thoroughly consolidated: he, however, objected to a German's leg which was protruding out of parapet, and I was told to have it buried forthwith by the C.O. I called Finnegan, and told him to remove the offending limb. As it would have meant pulling down the whole parapet to bury it, he took up a shovel and slashed at it with the sharp edge of the tool. After some hard bangs, he managed to sever the limb. I had turned away and was standing in the next fire bay, when I overheard Finnegan remarking to another man: "And what the bloody hell will I hang me equipment on now?"

'Three men of the Machine-Gun Section were wounded. We found a private of the York and Lancasters wounded and in a dying condition in a dug-out near the culvert, he appeared to have been there for days without any help. I had No. II Platoon carrying up bombs all night to the front line.

'17th August

'On duty all morning in the advanced trench. The C.O. brought the C.O. of the North Staffords, Lieut.-Colonel de Falbe, up to look round the line. He gave me orders about burying some dead. In a hollow he had discovered three un-buried. This was a sad sight, as the trio consisted of a patient lying on a stretcher and the two stretcher bearers lying across him, with the slings of the stretcher still across their shoulders. All had been knocked out by the same shell.

'We were only shelled in the support trench and at Railway Wood. At 10 p.m. we were relieved by the 1st North Staffords, and I handed over my line with its flank in the air joyfully! After relief we did not return to billets, but found carrying parties for R.E. material to the Hooge crater. So back again we toiled along the Menin road in Indian file, with duckboards, stakes, planks, and sand-bags. To make matters worse, it was raining hard and very dark. It was a tedious job: fallen trees had to be negotiated and numerous shell-holes full of water had to be avoided. The enemy was sending up star-shells, and we had to halt until the flare fell and had burnt itself out. To have been seen by the enemy would have been fatal, as we were on the exposed Menin road, right away from cover of any description. We finished our work at 1.30 a.m. and moved off for Ypres in the dark, and in heavy rain.

'We had no guide to meet us in Ypres, and we wandered about near the Lille Gate. Sergt. Sullivan, the Provost Sergeant, heard me cursing, and came to my assistance, and showed me our billeting area.'

Appendix 5

EXTRACT FROM HAIG'S DIARY, 9TH OCTOBER 1915

'Lord Haldane came to lunch. Afterwards he came to my room and asked me to give him my views on the action of the Reserves, i.e. of the 21st and 24th Divs. during the 25th and 26th September. He said that feelings were so strong on the subject in England that he had come to France in the hope of arriving at the truth. I gave him all the facts. The main criticism to my mind is the fact that the Reserves were not at hand when wanted. The causes for this seems to be:

'1. Neither the C-in-C nor his staff fully realized at the beginning (in spite of my letters and remarks) the necessity for reserves being close up before the action began.

'2. The two divisions were billeted in depth a long distance from where they would be wanted, and no attempt was made to concentrate them before the battle began.

'3. When the course of the fight showed that reserves were wanted at once to exploit the VICTORY, the two divns. were hurried forward without full consideration for their food, etc., with the result that the troops arrived worn out at the point of attack and unfit for battle.

'4. But the 21st and 24th Divns. having only recently arrived in France, with staffs and commanders inexperienced in war, should not have been detailed for this work. It was courting disaster to employ them at once in fighting of this nature. There were other divisions available as shown by the fact that they arrived three days later upon the battlefield, namely, the 28th Divn., the 12th Divn. and the Guards Divn.

'I also felt it my duty to tell Lord Haldane that the arrangements for the supreme command during the battle were not satisfactory. Sir John French was at Philomel (near Lillers) twenty-five miles nearly from his C.G.S. who was at St. Omer with G.H.Q. Many of us felt that if these conditions continued it would be difficult ever to win! Lord Haldane said that he was very glad to have had this talk with me, and seemed much impressed with the serious opinion which I had expressed to him.

'(NOTE: In spite of these views I expressed, as given above, to Lord Haldane, the latter went back to England and stated that no blame for failure could be attached to Sir John French.)'[1]

1. This comment was added subsequently by Haig.

Appendix 6

HAIG'S RECOMMENDATION TO THE CABINET, 24TH NOVEMBER 1915

'At 6.30 p.m. I attended at the Colonial office by appointment and saw Mr. Bonar Law. The main points I urged were:

'1. The immediate removal of the Imperial General Staff (with Sir Wm. Robertson as C.I.G.S.) to Horse Guards, so as to be free from War Office routine and questions of administration, General Staff to lay down the size of the Army required and how it is to be employed.

'2. The formation of only one class of Army instead of three as at present (Territorials, K's and Regular Armies).

'3. The Divisions to be of similar establishment throughout the Field Force.

'4. Units at the front to be maintained at full strength. My Army alone is 21,000 of all ranks deficient.

'I did not mention anything about Sir J. French. Mr. B. Law stated how Lord K had misled the Govt and wondered what appointment could be found for him to remove him from London!'

AUTHOR'S NOTE: One of the conditions that Haig shared with Sir John French was his apprehension that Lord Kitchener might be going to put 'his' armies to a special and personal use—probably in furtherance of K's strategic concepts in the Near East. Hence Haig's anxiety to take advantage of any weakness in Kitchener's position to prise away from him the I.G.S. and the administrative control of the Army.

The reference to French is of interest only as suggesting that the subject probably was, in fact, discussed. Bonar Law's indiscretion coming from one habitually so reticent is yet further evidence of Cabinet unity on this one question, if on no other—namely, how to get rid of Lord 'K'?

Bibliography

Atkinson, C. T., *The Devonshire Regiment 1914–1918* (Simpkin Marshall, 1926).

Beaverbrook, Lord, *Politicians and the War*, Vol. I (Thornton Butterworth, 1928).

Bickersteth, J. B., *History of the 6th Cavalry Brigade, 1914–1919* (Baynard Press, 1919).

Black Watch, The, *Regimental History, 1914–1919* (Ed. A. G. Warehope, Medici Society, 1925).

Blake, R. N. W. (Ed.), *The Private Papers of Douglas Haig* (Eyre and Spottiswoode, 1952).

Callwell, Sir E., *Memoirs of Sir Henry Wilson* (Cassell, 1927).

Charteris, Brigadier-General, *Earl Haig* (Cassell, 1929); *G.H.Q.* (Cassell, 1931).

Churchill, W. S., *World Crisis.* (Thornton Butterworth, Vols. I and II, 1923; Vol. III, 1927; Vol. IV, 1929; Vol. V, 1931).

Duff Cooper, *Haig* (Faber, 1935).

Edmonds, Sir J. E., *Military Operations in France and Belgium* (Macmillan, 1928); 'The Reserves at Loos' (In *Journal of the Royal United Service Institution*, Vol. LXXXI, 1936).

Ewing, J., *The Royal Scots, 1914–1919*, 2 Vols. (Oliver and Boyd, Edinburgh, 1925).

Falkenhayn, Field Marshal von, *General Headquarters and its Critical Decisions* (English translation, H.M.S.O., 1919).

French, E. G., *The Life of Field Marshal Sir John French, First Earl of Ypres, K.P., G.C.B., O.M., G.C.V.O., K.C.M.G.* (Cassell, 1931).

French, J. D. P., 1st Earl of Ypres, *1914* (Constable, 1919).

George, D. L., *War Memoirs* (Nicholson and Watson, 1933–6).

Hitchcock, F. C., *Stand To: A diary of the trenches, 1915–1918* (Hurst and Blackett, 1937).

Liddell Hart, B. H., *Reputations* (Murray, 1928); *History of the*

Great War, 1914–1918 (Faber and Faber, 1936); *Foch, The Man of Orleans* (Cassell, 1934).

Magnus, Philip, *Kitchener: Portrait of an Imperialist* (Murray, 1958).

Poincaré, R., *Au Service de la France*, Vols. VI, 1930; VII, 1931 (Plon, Paris, 1930–1); *The Memoirs of Raymond Poincaré 1915* (Translated and adapted by Sir George Arthur, Heinemann, 1930).

S . . . H. L., 'Loos' (In *The London Scottish Regimental Gazette*, Vol. XXV, No. 293, 1920).

Spears, Brigadier Sir Edward, *Prelude to Victory* (Cape, 1939).

Wynne, G. C., 'The Affair of the 21st and 24th Divisions at Loos, 26th September 1915' (In *The Fighting Forces*, Vol. XI, 1934).

Index

General: Officers are referred to by the rank which they bore at the time of the operations. Regiments are listed, but not by separate battalions.

BRITANNIA A History of Roman Britain by Sheppard Frere

'Brilliant . . . An integrated commentary and comprehensive judgement on the whole Romano-British scene.' *The Times Literary Supplement*

WEEK-END WODEHOUSE Introduction by Hilaire Belloc

'A peerless collection.' Max Hastings, *Sunday Times*

THE FACE OF BATTLE by John Keegan

'This without any doubt is one of the half-dozen best books on warfare to appear in the English language since the end of the Second World War.' Michael Howard, *Sunday Times*

ALAN MOOREHEAD by Tom Pocock

'Pocock's biography is excellent . . . it would be difficult to think of a better guide to the life of a Second World War correspondent.' Frank McLynn, *Sunday Telegraph*

THE CONTROL OF NATURE by John McPhee

'This splendid book describes three monumental acts of defiance against Mother Nature . . . The human drama is almost as breathtaking as the scale of these enterprises.' *Observer*

ARIEL A Shelley Romance by André Maurois

'An historic landmark in modern literary biography, as fine as any miniature produced by Lytton Strachey or Harold Nicolson.' Richard Holmes

COMPLETE VERSE by Hilaire Belloc; introduction by A. N. Wilson

'The verses sing a multitude of memories. The wonder is, in finding them all collected, how profuse and pure a genius is here displayed.' Evelyn Waugh, *Spectator*

CROSSMAN The Pursuit of Power by Anthony Howard

'Written from a background of great political knowledge and with shafts of penetrating insight . . . I doubt if it will be quickly, if ever, superseded.' Roy Jenkins, *Sunday Times*

GEORGIAN LONDON by John Summerson; a new illustrated and revised edition

'The scintillating text of this provocative classic was never matched by worthy pictures. Now, with a complete, handsome revamp, and with an Epilogue bringing the story up to date, the whole fine book at last truly reflects the perfection of the buildings.' Graham Hughes, *Arts Review*

SELF PORTRAIT WITH FRIENDS The Selected Diaries of Cecil Beaton Edited by Richard Buckle

'His book shows him to be sharper and more sardonic with his pen than with his camera – no less talented a portraitist, but an exacter and sometimes a crueller one. What this gifted, witty, sensitive creature writes best about is people. This is a minor masterpiece of wit and observation.' Cyril Ray, *Daily Mail*

F.E. SMITH First Earl of Birkenhead by John Campbell

'A triumph of scholarship, judgement, lucidity and art . . . Like its subject, John Campbell's book is leisurely, feline and very, very clever.' Roy Foster, *Guardian*

A VERY CLOSE CONSPIRACY Vanessa Bell and Virginia Woolf by Jane Dunn

'This is not only an important book in its own field, the triumphant outcome of years of loving concentration, but a book of rare discernment and imagination. It is one of the few books which allows me to believe that the author actually knew both my mother and my aunt, and I welcome it as deepening my understanding both of them and their relation to Bloomsbury.' Angelica Garnett

THE POUND ERA by Hugh Kenner

'Not so much a book as a library, or better, a new kind of book in which biography, history and the analysis of literature are harmoniously articulated . . . For the student of modern letters it is a treasure, for the general reader it is one of the most interesting books he will ever pick up in a lifetime of reading.' Guy Davenport, *National Review*